Soil, Plant, Water Analysis and Nutrient Management in Vegetable Crops

NIPA® GENX ELECTRONIC RESOURCES & SOLUTIONS P. LTD.
New Delhi-110 034

Soil, Plant, Water Analysis and Nutrient Management in Vegetable Crops

Pankaj Kumar
Rajaram Choyal
M.K. Jatav
R.C. Balai

NIPA® GENX ELECTRONIC RESOURCES & SOLUTIONS P. LTD.
New Delhi-110 034

NIPA® GENX ELECTRONIC
RESOURCES & SOLUTIONS P. LTD.

101,103, Vikas Surya Plaza, CU Block
L.S.C. Market, Pitam Pura, New Delhi-110 034
Ph : +91 11 27341616, 27341717, 27341718
E-mail: newindiapublishingagency@gmail.com
Web: www.nipabooks.com

For customer assistance, please contact
Phone: + 91-11-27 34 17 17
Fax: + 91-11- 27 34 16 16
E-Mail: feedbacks@nipabooks.com

ISBN: 978-93-58870-10-7

Composed and Designed by NIPA®.

Prof. Bhagirath
Vice-chancellor
MGSU, Bikaner
Maharaja Ganga Singh University
Bikaner (Raj.) India

Foreword

The rate of growth of agriculture is not commensurating with the growth of economy in recent years due to stagnating productivity of vegetable crops. In order to promote sustainable vegetable for the generations to come we must aim to maintain or improve soil health and avoid damage to the environment besides improving crop yield, quality and farmers' income. Fertilizers played very important role in realizing the role of green revolution. However, in the last few decades due to use of high yielding varieties, exploitative and intensive agriculture, indiscriminate use of high analysis chemical fertilizers has increased without knowing the availability of nutrients present in the soil. This has resulted into multiple nutrient deficiencies besides being nutrient imbalances in soils at several part/locations of the country. These situations have given rise to fatiguing effect on soil productivity and fertilizer use efficiency which has resulted into stagnation of productivity of major field crops. The balanced use of plant nutrients is required to correct nutrient deficiency improve soil fertility, increase nutrient and water use efficiency, enhance crop yield and quality and farmers income along with protection of environmental quality. To reap the benefits of good quality seed and better agronomic practices also it is important to have balanced use of plant nutrients.

Vegetables are short duration and nutrient exhaustive crop with poor nutrient efficiency due to much more sensitive to soil available as well as applied nutrients than major field crop. Soil, plant and water tests are the most important tools to practice judicious use of nutrient management in vegetable crops. Through soil testing farmers can know how much and what kind of fertilizer to use in a particular vegetable crop. A further refinement in fertilizer dose is possible on the basis of type of crop's variety, water availability and its quality, availability of organic manures, crop residues, biofertilizers, *etc.* The rapid tissue tests may help in real-time modification in nutrients to be applied.

I hope, this book on "Soil, Plant, Water Analysis and Nutrient Management in Vegetable Crops" will certainly help the field functionaries in guiding the farmers for adoption of soil and plant tests for judicious use of fertilizers and manures in vegetables production in the years to come.

-sd/-
Prof. Bhagirath

Preface

Food grain prices are soaring world wide due to fierce competition for reduced supplies of rice, wheat and maize as a result of their stagnant production. The vegetables are being seen as a vital food security crop and a substitute for costly food grains. However, the productivity of vegetables have also not witnessed an encouraging trend in last half a decade in our country. One of the most important reasons behind yield stagnation can be attributed to the land degradation and soil diminishing capacity to supply nutrients at the required rate due to over mining of essential nutrients under exploitative and intensive agriculture over the years. In addition, indiscriminate and imbalance use of fertilizers and lesser use of organic manures have also badly affected the soil health.

To restore back the soil health and break the yield plateau after relieving it from nutrient stresses, fertilizers and manures/bio-fertilizers have to be applied in a scientific manner taking consideration of the soil available nutrients as well as removal by crops. Soil, plant and water tests are the scientific way to ascertaining the soil health water quality for irrigation and balance nutrition to the crop. The vegetables being short duration and high photosynthetic efficient crop needs nutrient supply at very high rate as compared to other field crops. This is the reason why soil is rated sufficient in available nutrient for cereals may be deficient for vegetables. Therefore, soil and plant tests for this crop needs special care in deciding critical ranges/stages and balanced fertilizations.

Manual on soil, plant and water analysis and nutrient management in vegetables crop including development of methods, calibration and interpretation of results of soil, plant and water tests has been compiled in this book for ready use by vegetables growers/workers.

Authors

Contents

Appendices

1

Soil and Plant Tests - A Step Towards Judicious Use of Nutrients in Vegetables

Introduction

Many of the globe's major food crops *viz.,* wheat, rice and vegetables have reached "yield stagnation" or "yields plateau". This can be attributed to the land degradation and soil diminishing capacity to supply nutrients at the required rate due to over mining of essential nutrients. Further, imbalance use of fertilizers and lesser use of organic manures over the years have badly affected the soil health. To bring back life to the soil and relieve it from nutrient stress; fertilizers and manures/bio-fertilizers have to be applied in a scientific manner. At the same time, their indiscriminate use can result either in their under utilization and toxicity or leaching to the lower horizons, leading to underground water and create pollution problems. The soil is also the main source of sixteen mineral elements identifies in view of their importance for plant growth. **These are N, P, K, S, Ca, Mg, Na, Fe, Mn, Zn, Cu, Mo and Cl** (Tisdale *et al.,* 1985). Nitrogen, potassium and phosphorus have been designated as major nutrient elements in order of their removal by the plant; sulphur, calcium and magnesium as secondary nutrients while zinc, copper, iron, manganese and molybdenum are called micro nutrients as these are taken by plant in traces.

Nutrients need of vegetables depends upon the yield response to applied nutrients from fertilizers, agro-climatic conditions, variety and nutrient availability from the soil pool. The nutrient status of the soil further depends upon the quantity of total and available nutrients present in the soil pool. The soil tests helps in selecting the proper manuring schedule for various crops in a sequence by knowing about the nutrient build up or depletion in the soil profile. The adequacy of soil testing methods to correctly predict the fertilizer needs varies with the soil type, climate/ soil moisture and crop species. The rainfall also influences the efficiency of soil testing methods, especially nitrogen and potassium when it becomes imperative to have tissue analysis for modify the fertilizer recommendations. Plant tests also help to identify the nutrient deficiency in standing crop and in the event of deficiency of a particular nutrient helps in taking corrective measures (Sud *et al.*, 2008).

Soil testing

The practice of soil testing in India dates back to 1956-57 when 24 soil testing laboratories were established under Indo-US collaboration. The objective was to test the soil samples from cultivator's fields for pH and organic matter and providing valuable information regarding their soil productivity and for making fertilizer recommendations. This had played a major role in ensuring green revolution in the country (Ghosh and Hassan, 1979, 1980). Today with a net work of over 400 soil testing laboratories and opening of Krishi Vigyan Kendras at not only used district level in the country, concept of soil testing has undergone tremendous change. Now it is as tool for assessing soil fertility but also for getting information on the build up or depletion of soil nutrient pool, soil physical and biological properties. Thus it can play a vital role in economic and judicious use of manure and fertilizers. Several methods are available for assessing the suitability of soil for vegetables cultivation. In vegetables which is shallow rooted crop and has to depend more on the nutrients available near root zone in the top soil 30 to 40 cm layer for its growth. The soil testing has been mainly employed with the following objectives:

- Assessing the suitability of various soils for vegetables cultivation.
- Evaluating soil fertility status of vegetables growing soils and interpretation of soil test values.
- Fertilizer recommendation on the basis of soil tests.
- Judging the suitability of manuring schedule for the cropping systems.
- Fertility mapping of vegetables soils.

Collection and preparation of soil samples for analysis

The importance of having a true representative sample can be very well realized from the fact that only a minute fraction of huge soil mass of the field is actually used for the analysis in the laboratory to find out the quantity of essential nutrients available to plants and other relevant physical and chemical characteristic. While collecting soil samples the following aspects should be carefully considered.

The soil sample collected should be representative of the area sampled. A field can be treated as a single sampling unit only if it is appreciably uniform in all respects. Variation in slope, colour, texture, crop growth and management should be taken into account and separate sets of composite samples need to be collected from each of such area. Recently fertilized plots, bunds, channels, marshy tracts and sports near trees, wells, compost, or FYM/piles or other non representation locations must be carefully avoided during sampling. Soil sampling should be done well in advancer and not just before sowing. It has been observed that nutrient removal by a crop is almost completed after flowering stage and the nutrient status of soil does not alter. Therefore the soil sampling can be done in standing crop for the next crop. In general *Rabi* areas can be sampled after March and *Kharif* after September.

Ten to fifteen sub samples of approximately equal weight should be drawn in a zig zag pattern from an apparently homogenous plot. However, where the area with in the field looks different in appearance, topography, divided the field into parts and take sample of each portion (constituting a sample unit) separately. There is no rigidity about the size of the sampling unit. An experimenter in soil fertility has usually to sample individual plots in his experimental field. In case of saline-alkali soils, salt crust on the soil surface should be sampled separately and the depth of sampling is recorded.

Sampling procedure

To obtain a composite sample, small portions of soil are to be collected upto the desired depth. Generally the samples may be drawn to the plough depth (0 – 15 cm) by means of suitable sampling tools from at least 10 – 15 spots after scrapping off the surface litter, if any. However, some workers prefer a 30 cm depth. In pedological studies and in certain soil fertility investigations the samples are taken upto a depth of 180 cm or even deeper. When crops are planted in rows samples can be drawn in between the lines, if not recently fertilized (Fig 1).

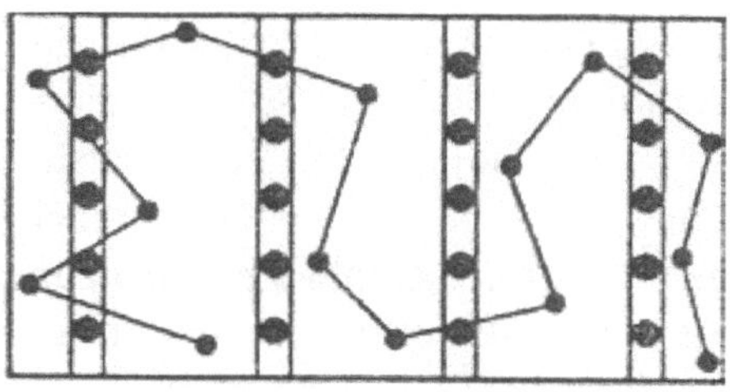

Fig. 1 : From Fields having standing crops collect soil from several spots in between the rows to prepare a composite sample

For sampling loose and moist soil, the tube auger, spade or khurpi is quite satisfactory (Fig. 2). A screw type auger is more convenient on hard/dry soils while post hole augar is useful for sampling excessive wet areas/ rice fields. If a spade or khurpi is used, a V shape cut may be made up to the plough layer and a uniform 1.5 cm thick slice is taken out.

The soil collected in this manner should be thoroughly powdered and mixed by hand on a clean piece of cloth or polythene sheet or thick paper. The bulk is reduced by quartering and about 500 g of the composite sample is retained. The soil must be quickly dried in shade at room temperature and put in cloth or polythene bags with suitable description and identification marks. Beside this cropping history of the area proposed cropping sequence should be mentioned.

The fertilizer application history and the results obtained should also be mentioned for prefect fertilizer recommendations (Sud *et al.,* 2008).

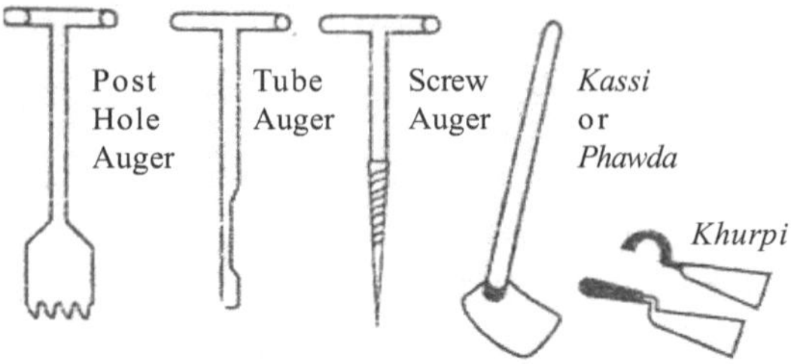

Fig. 2 : Use appropriate sampling tools. a Khurpi or tube auger proves more convenient for sampling from plough layer

Preparation of sample for analysis

Drying: Wet soil samples can not be stored as changes may occur in the chemical nature of c certain ions and organic matter. Samples are generally air dried (25–35°C), relative humidity (20-60%) and stored. For certain determinations such as ammonium and nitrate N, exchangeable K, acid extractable P and ferrous iron fresh samples from the field without any drying are required. Results of soil analysis are expressed on oven dry weight basis. This necessitation determination of moisture percentage by drying a small in an oven at 105°C for 2 hours.

Sieving: Field moist samples prior to drying can be made to pass though a 6 mm sieve (about 4 mesh per inch) by rubbing with fingers. The practice seems of much advantage in case of heavy soils. Soils in the right moisture condition can even be passed through a 2 mm sieve (about 10 meshes per inch). The common practice of sieving a portion of the gross sample through a 2 mm sieve and discarding the rest is undesirable as it increases the concentration of most of the elements involved in soil fertility. When the gravels in the soil exceeds 2% limit over a 2 mm sieve, their exact percentage should be recorded.

Grinding: A roller, rubber pestle in an agate mortar, or a motorized grinder is commonly employed. Crushing of the gravel or primary sand particles should be avoided, for heavy soils, as indicated earlier, it is better to pass these through a 2 mm sieve before allowing them to get completely air dried.

Mixing: Sample should be thoroughly mixed by rolling procedure. Place the dried, ground and sieve sample on piece of cloth. Grasp the opposite corners and then holding one corner down pull the other corner across the sample. Now the process is repeated in the reverse direction. Use the other two corners and roll the soil from one corner to another repeatedly. Continue this until thorough mixing is assured (Fig. 3).

Fig. 3 : Collect the soil at one clean place. Mix thoroughly by hand. Spread and make four quarters. Discard the two opposite ones. Remix the remaining two quarters. Repeat the process to reduce the quantity to about 500 g.

Storing: Store the soil in paper carton (soil sample box) using a polythene bag as an inner lining. Label the carton giving cultivators/experimenters name, plot number, date of sampling and initials (Fig. 4).

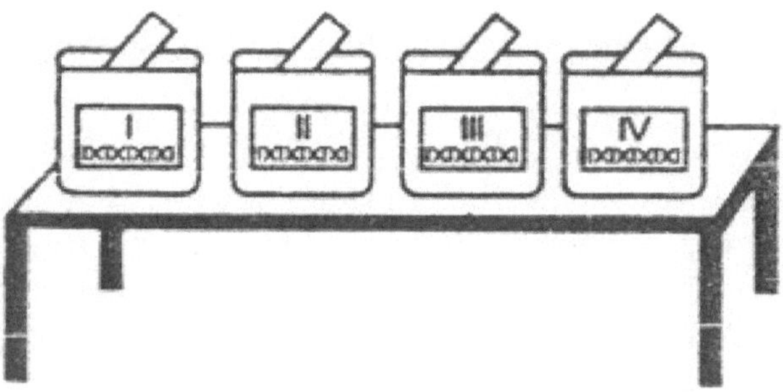

Fig. 4 : Transfer the soil to clean bag. Place one table inside the bag and another pasted outside, indicating sample no., name of owner, depth of sampling, Identification mark and send to soil testing laboratory

Physical tests: Soil tests for pH, texture, pore space bulk density *etc.* can determine its suitability of soil for vegetables cultivation. The foremost task is to test the soil for physical properties *i.e.* pH, texture, water holding capacity, bulk density, pore space and electrical conductivity as these parameters exert a great influence on nutrient availability to vegetables (Sud et al., 2008).

Soil pH

Vegetables can be grown in wide range of soils from acidic (pH 4.5) to alkaline (pH 8.5), however, the neutral soils are preferred. Soil pH (activity of hydrogen ions) has a direct influence on biological and metabolic reactions in the soil. At the same time, it also affects nutrient availability to the plants. High and low pH hinders vegetables plant growth as it affects the absorption of cations and anions by the plant roots (Fig. 5).

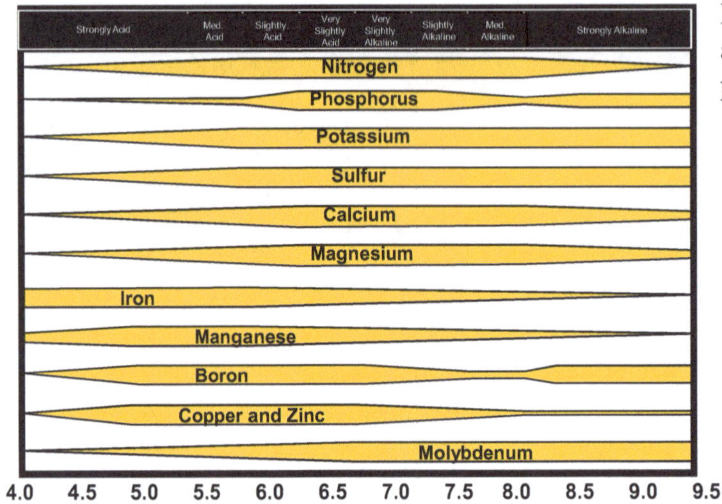

Fig. 5 : pH v/s availability of plant nutrients

Estimation of soil reaction (pH) (Jackson, 1973)

Soil pH has been defined as negative logarithm of the hydrogen ion activity.

$$pH = \log_{10} \frac{1}{a+} = - \log 10\ a^{H+}$$

Soil pH is measured by pH meter containing glass and reference electrode and marked pH scale from 0 – 14. The mid point 7.0 of this scale is neutral, below this denotes acidity and above this denotes alkalinity (Fig. 5). pH meter is standardized with the help of buffer solutions of known pH and then the pH of the solution is determined.

Apparatus

pH meter, vacuum pump, beaker, pipette, glass rod, china dish, spatula etc.

Reagents

Buffer solutions of pH 4.0, 7.0 and 9.2: One buffer tablet of the respective pH is dissolved in water and the volume is made to 100 ml.

Procedure

1. Saturation Paste is prepared by adding water to the soil and mixing till it starts glistening and slides on spatula. This paste is transferred to the bunchner funnel of vaccume pump and saturated extract is obtained.
2. 1:2 soil water extract is prepared by taking 20 g of soil and 40 ml distilled water in 100 ml beaker. The suspension is shaken at regular intervals for half an hour.

3. pH meter is set at room temperature and calibrated by immersing the electrodes in different buffer solutions of ph 4.0, 7.0 and 9.2.
4. Take the beaker of saturation or soil extract and dip the electrodes in to it and note the pH reading.
5. After each determination the electrodes must be washed distilled water and wiped out by ordinary filter paper.

Texture: Soil texture is an important tool in judging the suitability of the soil for vegetables cultivation since vegetables yield, shape and size all are governed by good texture. In addition, texture is a guiding force in quantifying soil's nutrients and water supplying capacity to the vegetables plant. As stated above, soil particles *viz.,* clay, silt and sand (coarse and fine sand) are essential components of soil solid phase and their ratio provide a key in defining the soil structure. Clay, silt and organic matter are known to be store house of nutrients and water. For vegetables, sandy loam and loamy soils containing high amount of organic matter are considered best. Majority of vegetables growing soils *i.e.* alluvial soils from Indo-Gangetic area are however, light textured having low organic matter and low water retention capacity. On the other hand, heavy textured clay loam to clayey soils from Peninsular India suffers from water logging, cracking and obstructs the growth. Vegetables growing soils should have adequate depth so as to permit better root growth and to facilitate earthing up operations.

As specified by the United States Department of Agriculture (USDA), the soil texture can be broadly taken as the amount of sand (2.0-0.050 mm), silt (0.050-0.002 mm) and clay (< 0.002 mm) present in the given soil sample. There are two methods which are followed in soil testing laboratories (i) *Bouyoucos hydrometer* and (ii) pipette method (Piper, 1950). In the former, the amount of different size fractions is determined with a soil hydrometer designed to measure the amount in grams of particles in suspension and is based on Law of Stokes. All particles >0.050 mm (sand) settles out after 40 seconds while silt settles out after 2 hours and the clay remains in the suspension. In the pipette method, a pipette is used to siphone off an aliquot of the soil suspension at a predetermined time and depth, then dried and weighed. In both the methods, the organic matter if present in considerable amounts is firstly destroyed by treatment with H_2O_2. The soil is suspended in 1L of water in a glass cylinder using either NaOH or $NaPO_3$ (Sodium Meta phosphate or Calgon) so as to expedite dispersion. Final dispersion is effected by thorough shaking or mixing mechanically with a blender. The values are then taken on to the USDA triangular guide for soil textural classification to find out the soil texture class.

Water holding capacity: It has a direct relationship with soil organic matter and clay content. Soils with high water holding capacity are considered best as they can with hold water for a longer period for subsequent use by the vegetables roots. The soil moisture plays a pivotal role in hilly areas of north-western, eastern in Himalayas and Nilgiri hills in the south where vegetables crop is taken under rainfed condition. During early stages of crop growth, the plant emergence is more influenced by the *in situ* soil moisture. Like wise, light textured alluvial soils from Indo-Gangetic plains having low organic matter content, the water holding capacity is less. These soils require heavy dose of fertilizers and water for getting optimum yields.

Weighed amount of 100 g soil is spread over glass and slowly water in small increments is added from the 100 ml burret and mixed well till the soil is well knit into dough and surface shines. The total amount of water is calculated that gives water holding capacity of the soil.

Bulk density: Soil bulk density, a ratio of the mass of dry solid to the bulk volume of the soil (includes pore space and solids) varies with structural condition of the soil. In general sandy soils and clayey soils have higher bulk density than those having high amount of organic matter. In heavy clayey soils of peninsular India, swelling of soils on account of irrigation/rains also affects the bulk density of the soil. Soil pore space controls the movement of water and nutrients within soil profile and helps the roots to penetrate deep in the soil layers. These properties are sometimes modified to some extent by periodic application of organic manures/ green manuring and crop residues.

Estimation of electrical conductivity of soil, (Jackson, 1973)

Principle: A solution offers some resistance to the passage of electric current through it depending upon the concentration and type of ions present. Higher the salt content, lesser the resistance to the flow of current. The resistance ® is defined by Ohm's law as the ratio of electrical potential in volys (E) and strength of current in ampere(I).

$$E/I = \frac{\text{Volts}}{\text{Current}} = R \text{ in ohm}$$

Electrical conductivity or conductance is the reverse of resistance.

I/R = I/Ohm = mho (reverse of ohm)

(At present mhos/cm is expressed in terms of dS/m).

Apparatus: Conductivity meter and a conductivity cell with known cell constant, vaccum pump, beakers, glass rod, china dish, spatula etc.

Reagents

i) Saturated solution of calcium sulphate (Reagent quality).

ii) 0.01 N KCL solution (Dissolve 0.7456 g of Potassium chloride in distilled water and dilute to one litre).

Procedure

1. The saturation extract of the soil or 1:2 soil water suspensions may be prepared as per method given in the procedure of pH determination.
2. Start the conductivity meter and adjust the temperature at 25°C.
3. Check the conductivity meter and adjust the temperature solution (conductivity –2.2 dS/m at 25°C) or 0.01 N KCL solution (conductivity – 1.41 dS/m at 25°C) before proceeding for the samples.
4. Take the reading of the saturation extract by dipping the conductivity cell into it. This will give ECe.
5. The same soil suspension prepared for determination of pH may also be used for EC. After recording the soil pH, allow the soil suspension in the beaker to settle for 30 minutes. Dip the conductivity cell and note the reading of conductivity meter. There should be no air bubble in conductivity tube.
6. Wash the conductivity cell after each determination and wipe ordinary filter paper.

2

Available Nutrients for Plant Growth in Soils

Estimation of organic carbon in soil (Walkley and Black, 1934)

Principle: Soil organic carbon is oxidized to CO_2 by heat of dilution obtained when a known amount of soil is treated with excess of standard $K_2Cr_2O_7$ in the presence of concentrated H_2SO_4. The excess of $K_2Cr_2O_7$ not reduced by the organic matter is back titrated with standard ferrous ammonium sulphate in the presence of phosphoric acid or sodium fluoride using diphenyl indicator. At the point colour changes from violet to blue or bright green.

Apparatus: 500 ml conical flasks, Pipettes of 10 and 20 ml, Burette, Balance, Measuring cylinder, Asbestos sheet.

Reagents

1. **Standard N $K_2Cr_2O_7$ solution:** Exactly 49.04 g $K_2Cr_2O_7$ (Potassium Dichromate) is dissolved in water and solution is diluted to one litre.
2. **0.5 N Ferrous ammonium sulphate solution:** This is prepared by dissolving 196.1 g Fe $(NH_4)_2SO_4 . 6H_2O$ in 800 ml of distilled water containing 20 ml concentrated H_2SO_4 and dilute to 1 litre.
3. **Diphenyl amine indicator:** 0.5 g of diphenyl amine is disolved in 20 ml of water and 100 ml. of concentrated H_2SO_4
4. **Orthophosphoric acid** (85%) or sodium fluoride.
5. **Concentrated sulphuric acid** (sp. gr. 1.84)

Procedure

1. Take 1 g (fine to moderately fine textured) or 2 g (Medium to coarse textured soil passed through 0.2 mm sieve (80 mesh per inch) in 500 ml conical flash.
2. Add 10 ml of N $K_2Cr_2O_7$ solution and shake well.
3. Keep the flask on an asbestos sheet and then add 20 ml of conc. H_2SO_4 by swirling the flask during addition.

4. Keep the content of the flask room temperature for 30 minutes to complete the reaction.
5. Keep one blank without soil.
6. After 30 minutes and about 200 ml distilled water and 10 ml of H_2 PO_4 or 0.5 g sodium fluoride and shake the contents of the flasks vigorously.
7. Add. 10 drops of diphenyl amine indicator it will give a violet colour to the suspension.
8. Titrate with N/2 ferrous ammonium sulphate solution till the colour changes from violet to blue or bright green.
9. Note the volume of ferrous ammonim sulphate solution used.
10. Carry out the titration of blank in similar manner.

Observation and Calculations

1. Weight of the soil sample used = S.
2. Volume of N/2 ferrous ammonium sulphate solution used for the blank titration = x ml.
3. Volume of N/2 ferrous ammonium sulphate solution used for titrating the excess of potassium dichromate = y ml.
4. Volume of 1 N of $K_2Cr_2O_7$ used for the oxidation of carbon = (x-y)/2 ml (1 ml of 1 N of $K_2Cr_2O_7$ = 0.0023 g of organic carbon).
5. Pre cent of organic carbon in the soil = x-y/2 X 0.003 X 100/S
   S = Weight of soil sample taken
   Organic carbon in soil g / kg = per cent of O.C. X 100

Since organic matter, on an average, contains 58% carbon, the per cent organic matter can be obtained by multiplying the organic carbon percentage with 100/58 or 1.725 which is known as Bemmlen factor.

Estimation of available nitrogen in soils (Subbiah and Asija, 1956)

Principle:-A known weight of the soil is mixed with excess of alkaline $KMnO_4$ and distilled. Ammonia gas released by distillation is absorbed in a known volume of standard sulphuric acid excess of which is titrated with standard alkali using methyl red as an indicator.

Apparatus:-Distillation apparatus, distillation flask, glass beads, 250 ml beakers conical flask, burette and pipette etc.

Reagent

1. **0.32 % $KMnO_4$ solution:** Dissolve 3.2 g pure $KMnO_4$ in distilled water and dilute to one littre.

2. **2.5 % NaOH solution:** Dissolve 25 g NAOH in distilled water and dilute to one litre
3. **0.02 N H_2SO_4:** Dissolve 0.56 ml of concentrated H_2SO_4 and make the volume one litre than standerdise with 0.02 N Na_2CO_3 using phenolphthelin and methyl red indicator or 0.02 NaOH using phenolpthelin indicator.
4. **0.02 N NaOH:** Dissolve 0.8 g NaOH in distilled water and dilute to one litre.
5. Methyl red indicator.

Procedure

1. Transfer 20 g of soil in a distillation flask. Add 20 ml water and 100 ml of 0.32 per cent $KMnO_4$ solution.
2. Take 20 ml of 0.02 N H_2SO_4 in a 250 ml conical flask and add two or three drops of methyl red indicator. Put the flask below the receiver tube. Keep the tip of the receiver tube well dipped into the H_2SO_4 solution.
3. Add 100 ml of 2.5 % NaOH to the distillation flask containing soil + $KMnO_4$ solution and immediately connect the flask with the distillation apparatus.
4. Start heating the distillation flask. The free ammonia thus released will be absorbed in the H_2SO_4 solution. Continue the distillation till the evolution of ammonia from distillation flask takes place, test it by bringing a moist red litmus paper near the outlet of condenser, which will turn blue as long as ammonia is being evolved. No change in red colour litus paper shows the completion of the ammonia distillation.
5. Titrate the excess of H_2SO_4 with N/50 NaOH and note the volume.

Observation and calculation

1. Weight of soil = 20 g
2. Volume of N/50 H_2SO_4 taken (X) = 20 ml
3. Volume of N/50 NaOH used (Y) =y ml
4. Volume of N/50 H_2SO_4 used for NH_3 (X-Y) = ml
5. 1 ml of N/50 H_2SO_4 = 0.00028 g N
6. Per cent available nitrogen = (X-Y) x 0.00028 x 100/20
7. Available nitrogen in ppm = percentage x 10^4
8. Available nitrogen Kg/ha = (X-Y) x 0.00028 x 2240000 or

   = (X-Y) x 28 x 1.12 = (X-Y) x 31.36

Estimation of available phosphorus in soil (Olsen *et al.*, 1954)

Several methods are used for the determination of available phosphorus in soils. Generally two methods are common. Olsen's method (Olsen *et al.*, 1954) is used for neutral and alkaline soil's where as the Bray's method (Bray and Kurty, 1945) is used for acid soils.

Olsen's method

Principle: In Olsen's method available phosphorus in soil is extracted by shaking the soil with 0.5 N sodium bicarbonate solution adjusted at pH 8.5. Later on the method involved formation of chlorostannous reduced molybdophospheric blue colour.

Apparatus: Klett Summerson colorimeter or Spectonic-20, shaking machine, conical flasks, volumetric flasks, pipette, measuring cylinder, funnel etc.

Reagents

1. **0.5 N sodium bicarbonate**- Take accurately 42 g of $NaHCO_3$ and dissolve in distilled water and make the volume one litre. Adjust the pH of this solution to 8.5 pH with 1 N NaOH.
2. **Darco-G 60 or activated charcoal** (phosphorus free)
3. **Ammonium molybdate 1.5 %**- Dissolve 15 g of ammonium molybdate in 300 ml of warm distilled water (about 50°C). Filter the mixture, if necessary and allow it to cool. Then add 342 ml. of concentrated HCl gradually by stirring it. Dilute to one litre with distilled bottle.
4. **Stannous chloride (stock solution)**- Dissolve 10 g of $SnCl_2$ in 25 ml concentrated HCl and add a piece of tin then heat it till it becomes clear. Store this solution in umber coloured bottle.
5. **Working solution of phosphorus**- Take 0.5 ml stock solution of $SnCl_2$ and make 66 ml with distilled water
6. **Standard solution of phosphorus:** Take 0.430 g of A.R. KH_2PO_4 (oven dried at 40°C) and dissolve in about half a litre distilled water. Add to this 25 ml 7 N H_2SO_4 and make the volume 1 litre with distilled water. This gives 100 ppm P (100 mg P per ml.)

Standard Curve

Prepare a 2.5 ppm P solution by taking 2.5 ml of 100 ppm solution in 100 ml volumetric flask and make its volume with distilled water. To prepare the standard curve of different concentration of P, take 1, 2, 3, 4, 5 and 10 ml of 2.5 ppm solution in 25 ml volumetric flask and follow the procedure for development of colour as given for sample and make the volume 25 ml. This will contain 0.1, 0.2, 0.3, 0.4, 0.5 and 1.0 ppm P. Also prepare a blank without adding KH_2PO_4 solution. The readings are vtaken on calorimeter against 660 μ (Red filter). The curve is plotted taking the calorimeter reading on vertical axis and amount of P in ppm on horizontal axis.

Procedure

1. Take 5 g of soil into a 250 ml conical flask and add a litre amount of phosphorus free activated charcoal.
2. Add 100 ml of 0.5 N $NaHCO_3$ solution into the flask and shake the conical flask for 30 minutes on a mechanical shaker.
3. Filter the suspension through a Whatman No. 42 filter paper and take 5ml aliquot of the extract in a 25 ml volumetric flask.
4. Add 5 ml of ammonium molybdate solution and a litre quantity of distilled water and shakes the contents of the volumetric flask.
5. Add 1ml of working $SnCl_2$ solution in each 25 ml volumetric flask and make the volume upto 25 ml with distilled water and shake well.
6. Measure the transmittance of the solution preferably within 5 to 20 minutes after the addition of $SnCl_2$ solution on 660 mu in a spectrophotometer or using red filter in Klett Summerson colorimeter.
7. Simultaneously run a blank without soil following the same procedure and take the reading.

Observations and Calculation

Weight of soil taken = 5 g

Volume of 0.5 N $NaHCO_3$ prepared = 100 ml

Aliquot of extract take = 5 ml

Klett reading of soil sample = X

Klett reading of blank sample =Y

Actual reading of soil sample = X-Y

ppm of P obtain from curve of (X-Y) = Z

Calculation

$$\text{Available P (kg / ha)} = \frac{\text{Z + Total volume of extract}}{\text{Volume of a liquot}} \times \frac{2.24 \times 10^6}{\text{Weight of soil (g) X } 10^6}$$

$$= \frac{\text{Z X 100}}{5} \times \frac{2.24}{5}$$

$$= \text{ppm P (Z) x 8.96}$$

$$= P_2O_5 \text{ (kg/ha)} = \text{ppm P (Z) x 8.96 x 2.29}$$

Estimation of available Potassium and Sodium in soil (Metson, 1956)

Available K is determined by extracting the soil by shaking with N neutral ammonium acetate solution. The ammonium ions replace potassium ions absorbed on the soil colloids. The estimation of potassium in the extract is carried out with the help of flame photometer.

Apparatus: Flame photometer, shaking machine, conical flasks, volumetric flasks, pipette, beaker, funnel etc.

Reagents

1. **N neutral ammonium acetate solution**: Take 700 ml of distilled water in a one litre volumetric flask. Add 57 ml of glacial acetic acid (99.5%) and then add 69 ml of concentrated ammonium hydroxide. Dilute it to about 900 ml and adjust the pH at 7.0 by adding 3 N NH_4OH or 3 N CH_3COOH and make up to one litre. It can also be prepared by dissolving 77.04 g of ammonium acetate in 900 ml of distilled water and adjust of the pH at 7.0 by adding NH_4OH or CH_3COOH as required and make the volume one litre.
2. **Standard solution**: Dissolve 1.908 g AR grade KCl in distilled water and make the volume one litre. This solution contains 1 mg K/ml i.e. 1000 ppm K.
3. **Working solution of K**: Take 10 ml of 1000 ppm K solution in 100 ml volumetric flask and make the volume. This will contain 100 ppm K.

Procedure

1. Take 5 g soil in 150 ml conical flask.
2. Add 25 ml ammonium acetate solution and shake for 5 minutes on a shaker.
3. Filter the contents through a Whatman No. 1 filter paper and reject first few drops of the filtrate.
4. Feed the filtrate to flame photometer after necessary setting and calibration of the instrument.
5. Also run a blank without soil using the same procedure.
6. Preparation of standard

Preparation of Standard Curve: From the 100 ppm solution take 10, 15, 20, 25, 30 and 40 ml of K in each 100 ml volumetric flask and make the volume. This will give 10, 15, 20, 25, 30 and 40 ml of K. The conc. of K may be reduced according to the working range of flame photometer. The curve is obtained by plotting the readings of flame photometer against the various concentrations of K by setting the reading at zero for the blank.

Observation and Calculation

Weight of soil taken = 5 g

Volume of extract = 25 ml

Reading of flame photometer = X

ppm of available K from the curve = R

$$\text{ppm of available K kg / ha} = R \times \frac{\text{Volume of extract}}{\text{Weight of soil}} \times \frac{2.24 \text{ x } 10^6}{10^6}$$

$$= R \times \frac{Z \text{ X } 100}{5} \times \frac{2.24}{5}$$

$$= R \text{ x } 11.2$$

Available K_2O Kg/ha $= R \text{ x } 11.2 \text{ x } 1.23$

Exchangeable Sodium in Soil

Exchangeable Sodium is determined in soil by shaking it with neutral ammonium acetate solution as described for determination of potassium in soil. 1000 ppm standard solution of sodium is prepared by dissolving 2.541 g of NaCl in 1 litre of distilled water. The same procedure of potassium determination is adopted for sodium determination is soil.

Estimation of calcium and magnesium in soil (Richards, 1954)

Principle: Calcium and magnesium react with Ethylene diaminetetra acetic acid (versinate) at different pH values and form stable complexes. Sn, Cl, Zn and Fe may interfere in the determination of Ca and Mn in magnesium. This interference may be prevented by using carbamate. In general a small amount of interfering elements has been found in soil extracts. In Ca estimation a known amount of soil extract is tirated with standard EDTA solution (0.01 N) using murexide indicator in presence of NaOH at pH 12, the colour changes at end point from orange red to purple, while for Ca + Mg Eriochrome black T indicator is used in the presence of $NH_4OH + NH_4Cl$ buffer at pH 10, the colour changes at end point from wine red to blue. The whole of Ca or Mg forms complex with EDTA.

Apparatus: Burette, pipettes, beakers, funnel, porcelin dish, measuring cylinder and glass rod etc.

Reagents

1. **Standard 0.01 N calcium solution:** Take 0.5 g of AR grade calcium carbonate and dissolve it in about 10 ml of 3 N HCl. Boil it to expel CO_2 and

then transfer in to one litre of volumetric flask and make the volume with distilled water.

2. **EDTA solution:** Weigh 2 g of sodium salt of EDTA, Dissolve in water in beaker and then transfer to one litre volumetric flask and make the volume with distilled water.
3. **Murexide indicator:** Take 0.2 g of ammonium purpurate and mix it with 40 g of potassium sulphate, mix these well in a pestle and mortar. It is stored in powder from because in solution it gets oxidized.
4. **Sodium hydroxide 4 N solution:** This solution is prepared by dissolving 160 g of NaOH into one litre of distilled water.
5. **Ammonium chloride ammonium hydroxide buffer:** Take 67.5 g of ammonium chloride and dissolve in 570 ml of conc. ammonium hydroxide and make it to 1 litre.
6. **Eriochrome Black T indictor:** In a beaker take 100 ml of ethyl alcohol and dissolve 4.5 g of hydroxylamine hydrochloride in it. Add 0.5 g of Eriochloride Black T indicator powder and prepare the solution. Hydroxylamine hydrochloride reagent prevents the interference of Mn.
7. **Sodium diethyl dithiocarbamate** crystals or 2 per cent solution of sodium cyanide.

Procedure

a) Calcium

1. Take a suitable volume of soil extract or saturation extract with the help of pipette in a Porcelain dish, if the concentration of Ca is higher than 60me/l than extract may be diluted.
2. Add two to three crystals of carbamate and 5 ml of 4 N NaOH solution and shake with the help of glass rod.
3. Add 40-50 mg or a pinch of murexide indicator & shake, it will give red colour to the solution.
4. Titrate the contents of porcelain dish with 0.01 N EDTA solution till the colour changes from red to reddish violet or purple. It is advised to add EDTA drop by drop from burette. As the change in colour is not very sharp, the end point should be compared with a solution without Ca. if the solution is over titrated it can be back titrated with standard Ca solution and by subtracting it from the sample reading. Then exact volume used for Ca titration can be obtained.

b) Calcium + Magnesium

1. Pipette about 10-20 ml of soil extract containing not more than 0.1 me of Ca + Mg in a porcelain dish if the solution contain higher concentration then it should be diluted.
2. Add two to five crystals of carbamate and 5 ml of ammonium chloride ammonium hydroxide buffer and shake well with glass rod.
3. Add 3-4 drops Eriochrome Black T indicator, the colour of the solution become wine red.
4. Titrate the contents of porcelin dish with 0.01 N EDTA solution till the colour changes from wine red to blue and there should not remain any tinge of wine red colour.

Observation and Calculation

1. Weight of soil = 100 g.
2. Volume of extract = 500 or 1000 ml.
3. Volume of aliquot taken for Ca =
4. Volume of aliquot taken for Ca + Mg =
5. Volume of 0.01 N EDTA solution used for Ca titration = X ml.
6. Volume of 0.01 N EDTA solution used for Ca + Mg titration = Y ml.

Estimation of available sulphur in soil (Chesnin and Yien, 1957)

Principle: Soil is extracted using monocalcium phosphate solution. The extent of turbidity due to fine suspension of barium sulphate formed by the reaction of sulphate ions with barium chloride in solution can be accurately measured at low concentration by a Spectrophotometer or Colorimeter.

Apparatus: Colorimeter, shaker, conical flask, pipette, funnel, volumetric flask, measuring cylinder etc.

Reagents

1. **Extracting solution:** Dissolve 2.18 g of MCP i.e. $Ca(H_2PO_4)_2.2H_2O$ in distilled water and dilute to one litre
2. **Gum acacia solution (0.25%):** Dissolve 0.25 g of gum acacia in distilled water and dilute to 100 ml.
3. **Barium chloride (30-60 mesh crystal):** Grind $BaCl_2$ crystals in an ageta-morter until crystals pass through 30 mesh sieve and are caught on 60 mesh sieve.

4. **Standard sulphur solution (100 ppm S):** Dissolve 0.5434 g of the reagent grade K_2SO_4 in distilled water and dilute to one litre.

Preparation of Standard curve: Pipette out 0.25, 0.5, 1.0, 2.5, 5.0 and 7.5 ml of 100 ppm S solution in 25 ml volumetric flask, respectively. The final volume will give 0, 1, 2, 4, 10, 20 and 30 ppm sulphur, respectively. Add in every flask 10 ml of mono. Calcium phosphate extractant. Then follow the steps from 3 to 6 given in the procedure before taking reading.

Procedure

1. Take 20 g of air dried soil and transfer to a 250 ml flask or plastic bottle. Add 100 ml of the extracting solution. Shake for 30 minutes & filter the contents through Whatman 42 filter paper.
2. Transfer 20 ml of the aliquot of extract to a 25 ml volumetric flask.
3. Add 1 g of 40 to 60 mesh $BaCl_2$ crystals and shake for one minutes.
4. Then add 1 ml of 0.25% gum acacia solution.
5. Make up the volume to 25 ml and shake for one minute.
6. Take turbidity readings, within a period of 5 to 30 minutes using a colorimeter at wave.
7. Run a blank also.

Observation and Calculation

- Weight of soil = 20 g
- Volume of extract = 100 ml
- Volume of aliquot taken = 20 ml
- Final volume = 25 ml

$$\text{ppm of available S in soil} = R \times \frac{100}{20} \times \frac{25}{20}$$

(R= corresponding ppm from curve)

$$\text{Available S kg/ ha} = R \times 6.25 \times \frac{2.24 \text{ X } 10^6}{10^6}$$

$= R \text{ x } 6.25 \text{ x } 2.24$

Available $SO_4 - S$ kg/ ha = R x 6.25 x 2.24 x 3

Estimation of available iron, manganese copper and zinc in soil (Lindsay and Norwell, 1978)

Principle: Diethylene tetramine penta acetic acid (DTPA) solution helps in the determination of available Fe, Mn, Cu and Zn being a chelating agent, when the soil is shaken using a solution of DTPA, it combines with metal ions in the solution and form soluble complexes of Fe^{++}, Mn^{++}, Cu^{++} and Zn^{++}. $CaCl_2$ and triethanolamine (TEA) solution slightly raise and buffer the soil pH and further mitigates the effect of Ca^{++} and Mg^{++}. Ca^{++} and Zn^{++} cations equivalent to 10 times of their atomic weights which ranges from 550 to 650 ppm. Therefore, when DTPA present in a sizeable amount the nutrient cations are normally solubilized during extraction. From DTPA extract Fe^{++}, Mn^{++}, Cu^{++} and Zn^{++} are determined with the help of Atomic Absorption Spectrophotometer by using their respective hollow cathodes.

Apparatus: Atomic Absorption Spectrophotometer, shaking machine Centrifuge, beaker, pipette, volumetric flask, conical flask, funnel etc.

Reagents

1. DTPA 0.005 M Solution
2. TEA 0.1 M (AR or extra pure) solution
3. $CaCL_2.2H_2O$ (AR) 0.01 M
4. Dilute HCl AR diluted with double distilled water 1:1

The extracting reagent is prepared by taking 1.967 g of DTPA, 1.470 g $CaCl_2.2H_2O$ and 13.3 ml of TEA in 100 ml of glass or double distilled water and dilute approximately to 900 ml. Adjust the pH of the solution to 7.3 by adding 1:1 HCl with the help of pH paper. While stirring make the volume of the extracting solution to one litre. The solution remains stable for several months.

Preparation of 100 ppm standard solution

Element	Atomic weight	Micro- nutrient Salt	Molecular weight	Quantity of salt in g for 1 litre solution
Zn	65.38	$ZnSO_4.7H_2O$	287.56	0.4398
Mn	63.54	$CuSO_4.5H_2O$	249.69	0.3929
Cu	57.94	$MnCl_2.4H_2O$	197.69	0.3602
Fe	55.85	$FeSO_4.7\ H_2O$	278.02	0.4977

Procedure

- Weight 10 g of soil and transfer it to a 150 ml polythene bottle and add 20 ml DTPA extracting solution to it and stopper it well.
- Shake the contents of flask on electric shaker for 120 minutes and centrifuge or filter the contents through Whatman No. 42.

- Prepare a blank by adding all the solution except soil by shaking and filtration.
- Insert the hollow cathode of the micronutrient which is to be determined in AAS & start the AAS.
- Feed the blank solution of 0 ppm and adjust the zero in AAS then feed the lowest concentration standard and note the reading after that take the reading of standard in increasing order of concentration by running the blank in between two standards to ensure that readings has not changed from zero. Note the readings of each working standard.
- Absorbance verses concentration of standards is plotted on graph paper and prepare the standard curve.
- Feed the sample extract one by one into AAS and note the readings.
- Find out the concentration in ppm of the sample with the help of standard curve.

Observation and Calculation

Weight of soil = 10 g

Volume of extractant = 20 ml

Volume of extractant 20

$$\text{Dilution factor} = \frac{\text{Volume of extractant}}{\text{Weight of soil}} = \frac{20}{10} = 2$$

Weight of soil 10

ppm of micronutreint in soil = ppm of sample reading x 2

Estimation of available molybdenum in soil (Grigg, 1953)

Principle: when soil is treated with ammonium oxalate solution of ph 3.3 molybdate absorbed on soil colloids is presumably replaced by the oxalate ions. the exchange is made irreversible by the formation of molybdanum-exalic acid complexes $MoO_3C_2H_2O_4$ and $(MoO_3)_2\ C_2H_2O_4$ the dark green complex is formed by reacting toluene-3, 4-dithiol $(CH_3C_6H_3S_2)_3$ with Mo. the absorbance is measured in a spectrophotometer at 680nm.

Apparatus: Spectrophotometer or colorimeter, horizontal shaker, conical flask, water bath, muffle furnance, measuring cylinder, funnel, beaker, volumetric flask, conical flask, pipette.

Reagents

1. **50% Potassium iodide aquous solution:** Dissolve 50 g of KI in 100 ml of double distilled water.

2. **50% Ascrobic acid aquous solution:** Dissolve 50 g of ascorbic acid in 100 ml of DDW.
3. **10% Sodium Hydroxide solution:** Dissolve 10 g of NaoH in 100 ml if DDW.
4. **Thiourea aquous solution 10%:** Dissolve 10 g thiourea in 100 ml of distilled water and filter. Prepare fresh solution before use.
5. **Dithiol solution (Tolune-3, 4-Dithiol):** Weigh 1.0 g of AR grade melted dithiol 51⁰ C in a 250 ml glass beaker. Add 100 ml of the 10% NaOH solution and warm the contents up to 51°C with frequent stirring for 15 minutes. Add 1.8 ml of thioglycolic acid and store in a refrigerator.
6. **Tartaric acid solution 10%:** Dissolve 10 g in 100 ml of DDW.
7. **Ferrous ammonium sulphate solution:** Dissolve 63 g of the salt in about 500 ml of DDW and then make the volume to one litre.
8. **Extracting agent**: Dissolve 24.9 g of AR grade ammonium oxalate and 12.6 g oxalic acid in water and make the volume to one litre. Adjust the pH at 3.3 with dil.HCl/ dil. Ammonia solution.
9. **Standard stock solution (100 µg Mo mL^{-1}):** Dissolve 0.150 g of AR grade MoO_3 in 100 ml of 0.1 N NaOH, Make slightly acidic with dil. HCl and make the volume up to 1 litre.
10. **Working standard solution (1 µg Mo mL^{-1}):** Dilute 10 ml of the stock solution of 1 litre
11. **Iso-amyl acetate**
12. **Ethyl alcohol**

Procedure

1. Take 25 g soil in a 500 ml corning or pyrex glass conical flask and add 250 ml ammonium oxalate of pH 3.3.
2. Shake the contents of the flask in and end to end horizontal shaker for 10 hours and then filter through Whatman filter paper No. 50.
3. Take 200 ml of the filtrate in a 250 ml beaker and evaporate to dryness on a water bath.
4. The contents of the beaker are transferred t a silica dish and then heated at high temperature (500°C) in a furnace for 5 hours to destroy organic matter and oxalates.
5. After this contents are digested with 5 ml of diacid (HNO_3-$HCIO_4$ 4:1). Keep it over night, then treat with 10 ml of 4N H_2SO_4 and H_2O_2 every time taking to dryness.

6. Then add 10 ml of 0.1 N HCI and filter it through Whatman No. 50 in 100 ml volumetric flask. Wash the dish with another 10 ml of 0.1N HCI. Then wash the filter paper with redistilled water until the filtrate volume obtained is 100 ml.
7. Store the filtrate in corked plastic reagent bottles for estimation of available Mo.
8. Transfer the contents in 250 ml separating funnel and follow the process described above for the standard curve to develop the colour and take the reading of the sample at 680 ml in a Spectrophotometer.

Observation and Calculation

Weight of soil taken = 25 g

Volume of extract made = 250 ml

Volume of filtrate taken = 200 ml

Volume of extract prepared (After drying 200 ml filtrate) = 100 ml

Dilution factor – 250/25x100/200 = 5

ppm of Mo in soil = ppm of sample reading x Dilution factor

ppm of Mo in soil = ppm of Mo in sample x 5

Estimation of available boron in soil (Gupta, 1979)

Principle: Hot water soluble boron in soils is generally expressed as available boron. This is treated with buffer masking solution of ammonium acetate and EDTA di-sodium salt in glacial acetic acid and the colour id developed by azomethine H due to the formation of stable complex with H_3BO_3 at pH 5.1. Boron concentration is measured in a Spectrophotometer at 420 nm.

Apparatus: Conical flasks, volumetric flasks, pipette, beaker, polythene bottle, heater, funnel, colorimeter, hot plate and quartz flask etc.

Reagents

Buffer masking solution: Dissolve 250 g of ammonium acetate and 15 g of EDTA di-sodium salt in 400 ml of distilled water. Add 125 ml of glacial acetic acid and mix it well.

Azomethine-H: Dissolve 0.450 g of azomethine in 100 ml of 1% ascorbic acid solution. Fresh solution prepared is stored in a regrigerator.

Activated charcoal

Standard solution: To prepare 100 ppm boron solution dissolve 0.570 g boric acid HBO (AR) grade in double distilled water and make the volume 1.0 litre. Take 5 ml of 100 ppm B solution in 100 ml volumetric flask and dilute to the mark. This solution contains 5 ppm B.

Procedure

1. Take 25 g soil sample in a quartz flask. Add 50 ml of double distilled water and about 0.5 g of activated charcoal into it.
2. Boil the contents of the flask and filter immediately through filter paper Whatman No. 42.
3. Take 5 ml of the extract in a 25 ml volumetric flask. Add 4 ml of buffer masking solution and 4 ml of azomethine – H reagent. After this the contents are allowed ot stand for one hour for the development of colour and volume is made to the mark.
4. Insert a blue filter in a colorimeter or adjust the wavelength at 420 nm in a Spectrophotometer and start the instrument.
5. Take the readings of standard solution and make a standard curve by plotting concentration of B on X-axis and absorbance on Y-axis on a graph.
6. Take the readings of sample and calculate the concentrations of B in analysed sample from the standard curve.

Observation and Calculation

1. Weight of soil = 25 g
2. Volume of extractant = 50 ml
3. Volume of extract taken = 5 ml
4. Volume prerared = 25 ml

Dilution factor = Volume of extractant / wt. of soil X Volume prepared for colour development/aliquot taken = 50/25 X 25/5 =10

Boron (ppm) in soil sample = Concentration of B in analysed sample from standard curve X10

Lime requirement

Crop yields are normally high in soils with pH values between 6.0 and 7.5. Lime is added to raise the pH of acid soils, and the amount of lime required to raise the pH to an optimal level is called lime requirement. Various methods are available for determining the lime requirement. The methods discussed here are based on the use of a buffer solution, whose pH undergoes change when treated with acid soils. The pH of the buffer solution decreases gradually as H^+ ion concentration increases. When H^+ increases by 1 me in 100 ml of buffer solution, the pH value decrease by 0.1 unit.

Shoemaker method

For the method developed by Shoemaker, McLean and Pratt (1961)

Apparatus: a pH meter; some automatic pipettes (10 and 20 ml).

Reagents

To prepare the reagent (extractant buffer), dissolve 1.8 g of nitrophenol, 2.5 ml of triethanolamine, 3.0 g of potassium chromate (K_2CrO_4), 2.0 g of calcium acetate and 53.1 g of calcium chloride in 1 litre of water. Adjust the pH to 7.5 with NaOH.

Procedure

- Place 5.0 g of soil sample in a 50 ml beaker.
- Add 5 ml of distilled water and 10 ml of extractant buffer.
- Shake continuously for 10 minutes (or intermittently for 20 minutes), and read the pH of the soil buffer suspension with a glass electrode. The pH of the buffer solution is reduced, depending on the extent of soil acidity.
- For various levels of measured pH of soil buffer suspension, Table 1 shows the amount of lime required in terms of $CaCO_3$ in order to raise the soil pH to 6.0, 6.4 and 6.8. Lime requirement varies with the type of soils and its CEC.
- For practical purposes, the pH of acid soils should not be raised beyond 6.4–6.5.

Table 1 : Lime requirement for different pH targets

Measured pH of soil buffer suspension	Lime requirement as $CaCO_3$ for bringing soil pH to different levels		
	6.0 (t/ha)	6.4 (t/ha)	6.8 (t/ha)
6.7	2.43	2.92	3.40
6.6	3.40	4.13	4.62
6.5	4.37	5.35	6.07
6.4	5.59	6.56	7.53
6.3	6.65	7.78	8.99
6.2	7.52	8.93	10.21
6.1	8.50	10.21	11.66
6.0	9.48	11.42	13.12

Gypsum requirement

In the estimation of gypsum requirement of saline-sodic/sodic soils (Schoonover, 1952), the attempt is to measure the quantity of gypsum (calcium sulphate) required to replace the Na from the exchange complex. The Na so replaced with the Ca of the gypsum is removed through leaching of the soil. The soils treated with gypsum become dominated with Ca in the exchange complex.

When the Ca of the gypsum is exchanged with Na, there is a reduction in the Ca concentration in the solution. The quantity of Ca reduced is equivalent to the Ca exchanged with Na. It is equivalent to the gypsum requirement of the soil when "Ca" is expressed as $CaSO_4$.

Apparatus

Apparatus : A mechanical shaker; a burette (50 ml); some pipettes (100 ml and 5 ml).

Reagents

The reagents required are:

1. **Saturated gypsum (calcium sulphate) solution**: Add 5 g of chemically pure $CaSO_4 . 2H_2O$ to 1 litre of distilled water. Shake vigorously for 10 minutes using a mechanical shaker and filter through No. 1 filter paper.
2. **0.01N $CaCl_2$ solution**: Dissolve exactly 0.5 g of AR-grade $CaCO_3$ powder in about 10 ml of 1:3 diluted HCl. When completely dissolved, transfer to a 1-litre volumetric flask and dilute to the mark with distilled water. Do not use $CaCl_2$ salt as it is highly hygroscopic.
3. **0.01N versenate solution:** Dissolve 2.0 g of pure ethylenediamine tetraacetic acid (EDTA) – disodium salt and 0.05 g of magnesium chloride (AR-grade) in about 50 ml of water and dilute to 1 litre. Titrate a portion of this against 0.01N of $CaCl_2$ solution to standardize.
4. **Eriochrome Black T (EBT) indicator:** Dissolve 0.5 g of EBT dye and 4.5 g of hydroxylamine hydrochloride in 100 ml of 95 per cent ethanol. Store in a stoppered bottle or flask.
5. **Ammonium hydroxide-ammonium chloride buffer:** Dissolve 67.5 g of pure ammonium chloride in 570 ml of concentrated ammonium hydroxide and dilute to 1 litre. Adjust the pH to 10 using dilute HCl or dilute NH_4OH.

Procedure

- Weigh 5 g of air-dry soil in a 250-ml conical flask.
- Add 100 ml of the saturated gypsum solution. Firmly insert a rubber stopper, and shake for 5 minutes.

- Filter the contents through No. 1 filter paper.
- Transfer 5 ml of aliquot of the clear filtrate into a 100 or 150-ml porcelain dish.
- Add 1 ml of the ammonium hydroxide-ammonium chloride buffer solution and 2–3 drops of EBT indicator.
- Place 0.01N versenate solution in a 50-ml burette and titrate the contents in the dish until the wine red colour starts to change to sky blue.

Observation and calculation

Volume of versenate used = B.

Run a blank using 5 ml of saturated gypsum solution in place of sample aliquot. Volume of versenate solution used = A.

The gypsum requirement (in tonnes per hectare) is given by: $(A - B) \times N \times 382$; where:

A = millilitres of EDTA (versenate) used for blank titration;

B = millilitres of ETDA used for soil extract;

N = normality of EDTA solution.

3

Plant Sampling and Sample Preparation

For plant analysis to be meaningful, collection of particular plant part(tissue) at the right stage of growth is very important. A research worker in soil fertility is generally interested in finding out the nutrient uptake by crop at different stages. The stage at which nutrient uptake is to be studied will vary with different crops. One may also be interested in collecting a plant sample when some abnormality is observed in the plants and a nutrient deficiency or toxicity is suspected. In such cases no time should be lost in collecting the sample otherwise secondary reactions that often take place in the mineral content of plants will conceal the real cause of deficiency or toxicity. In such cases only the affected plants are to be sampled as compared to general sampling technique described below.

For research projects each plot receiving differential fertilizer or some other treatment is to be sampled separately. For general advisory purposes one sample would be good enough for the entire field (not exceeding 2 hectares). One should, however, take care that the sample comes from a field or a part there of having a uniform stand. If a field is patchy and differences in stand and plant growth are obvious, each such area should be sampled separately.

Sampling technique

The person collecting the sample should travel across the rows in the centre of the plot. In small plots every alternate row and in big plots very 5th to 50th row (depending upont he size of field) should be samples. For small experimental plots (100 sq m) about 10 plants would serve the purpose but for bigger fields about 25-50 plants should be collected. It is always advisable to collect a larger sample in the field and then reduce it in the laboratory as it helps in getting a representative sample. For working out the plant nutrient uptake by crop, entire plants are to be sampled. Different tissue should then be separated (their dry weight must be recorded) and analysed separately.

After sample collection, the fresh tissue should be decontaminated from dust and other foreign material by adopting the following procedure. Three plastic containers or glass beakers are taken in which 0.2% liquid detergent, N/10 HCI solution (8 ml conc. HCI/litre) & deionized or distilled water solutions are added separately.

The fresh tissue should be washed in sequence in detergent solution, dilute HCI and deionized water. The liquid detergent will remove waxy coating on leaf surface and any soil particles. N/10 HCI will remove metallic contaminats and deionized water will wash the previous two solutions. The extra moisture is wiped out, the sample is placed in new paper bags and dried in an oven at 70°C.

In case the sample site is away from laboratory which requires longer travel priod, the sample may be placed in polythene bags and transported in an ice chest. After drying the samples are kept in paper bags. In the laboratory, the samples are homogenized using a sample mill which should not give any metallic contamination. After grinding, the sample may be dried at 70°C before taking up weighing for digestion.

Estimation of nitrogen in plant (Snell and Snell, 1949)

Colorimetric method

Principle: Plant sample are digested with H_2SO_4 and then with H_2O_2 till it become colourless. The intensity of the colour developed by Nessler's reagent in the presence of NaOH and sodium silicates is measured colorimetrically.

Apparatus: Kjeldahl flask 100 ml, measuring cylinder, volumetric flask 50 and 100 ml, pipette, beaker, tilt measure, digestion assembly, colorimeter etc.

Reagents

1. **Sulphuric acid (concentrated)**
2. **Hydrogen peroxide 30%**
3. **Nessler reagent:** Dissolve 45.5 g mercuric iodide and 35.0 g potassium iodide in about 70 ml distilled water. Then dissolve 112 g potassium hydroxide in another beaker in 100 ml distilled water. Mix both the solution in one litre volumetric flask and add 600 ml of water. Shaker the contents of flask thoroughly, cool it and make the volume up to the mark. The solution is allowed to stand for few days and the clear supernated liquid is stored in to an umber coloured bottle for use. Working solution of Nessler reagent is prepared by diluting the stock solution four times i.e. by diluting 100 ml to 400 ml.
4. **Sodium silicate 10% solution:** Disolve 10 g of sodium silicate in 100 ml of water. Boil it till it becomes clear. This solution is used to prevent turbidity.
5. **Sodium hydroxide 10 % solution :** Dissolve 10 g sodium hydroxide pellets in 100 ml of distilled water. Freshly prepared solution is used to neutralize the excess acid.

6. **Nitrogen standard solution:** Dissolve 0.1179 g of $(NH_4)_2SO_4$ AR grade in water and make the volume one littre in volumetric flask. This solution contains 25 ppm N. For working standards take 0,1,2,3,4,5,6,7,8 ml of 25 ppm N solution in 50 ml volumentric flasks. Add 2.0 ml. 10 % NaOH and 1 ml 10 % sodium silicate solution in 50 ml volumetric flasks. Add some water and shake the contents. Now add 1.6 ml Nessler reagent slowly drop while shaking and make the volume. The flask will contain 0, 1.0, 1.5, 2.0, 2.5, 3.0, 3.5 and 4.0 ppm N, respectively.

Procedure

1. Take 0.1 g well ground and dried plant sample and transfer into a 100 ml dry Kjeldahl flask carefully so that it should not stick to the neck.
2. Add 2 ml concentrated H_2SO_4 to the Kjeldahl flask, mix the contents of the flask. Place on the digestion assembly and heat it till the plant material is digested.
3. Cool the flask and add 0.5 ml (1 drops) of 30% H_2O_2 and again heat the contents this till it becomes clear and colourless.
4. Transfer the digested contents of the Kjeldahl flask into a 100 ml volumetric flask by washing it 2-3 times with distilled water and make the volume.
5. Take 5 ml of digested solution into a 50 ml volumetric flask and add few ml of water. Then add 2 ml of 10% NaOH and 1 ml of 10% sodium silicate solutions and then add some water. Mix the contents and add 1.6 ml Nessler's reagent to the flask while shaking. After this, make its volume.
6. Take the reading of standard working solutions using blue filter or adjusting the spectrophotometer at a wave length of 420 nm. Plot the concentration of N on X-axis and the colorimeter reading on Y-axis and prepare a standard curve.
7. Now take the reading of plant sample and calculate the ppm N with the help of standard curve.

Observation and Calculation

1. Weight of plant sample = 0.1 g
2. Volume of digested material = 100 ml
3. Volume of extract taken = 5 ml
4. Final volume prepared = 50 ml
   Dilution factor = 100/0.1 X 50/5 = 10,000
   Concentration of N in ppm = ppm N from standard curve ® X 10,000
   %N in plant sample = R x 10,000/10,000 =R

Estimation of phosphorus in plant (Jackson, 1973)

Principle: Orthophosphates obtained in triacid digest react with ammonium molybdate ammonium vanadate in HNO_3 medium and gives a yellow colour complex. The colour develops in about 30 minutes and remains stable for 2 to 8 weeks. Intensity of colour is measured at 470 nm in a Spectrophotometer.

Apparatus: Spectrophotometer, volumetric flask, pipette, conical flasks, digestion flasks, 1 litre and 50 ml beakers.

Reagents

1. **Triacid:** First take 100 ml of HNO3 add 10 ml of concentrated H_2SO_4, mix it and then add 40 ml of 60% $HClO_4$ i.e. in the ratio of 10:1:4 of HNO_3: H_2SO_4: $HClO_4$.
2. **Ammonium molybdate ammonium vanadate solution:** Dissolve 22.5 g of $(NH_4)_6Mo_7O_{24}.4H_2O$ in 400 ml of distilled water in a beaker. Take 1.285 g of ammonium vanadate in another beaker, add 300 ml distilled water and boil it. Add ammonium vanadate solution to the ammonium molybdate solution and cool the contents. Add 250 ml of concentrated HNO_3 and dilute it to 1 litre.
3. **Phosphate standard solution:** Take 0.2195 g of AR grade KH_2PO_4 in a beaker and dissolve in distilled water, transfer the solution to a 1 litre volumetric flask and make up the volume . This solution contains 50 ppm of P.

Procedure

1. Take 0.5 to 2.0 g of dried and ground plant sample in a 500 ml conical flask and add 5 ml of concentrated HNO_3 for each 1 g of plant sample. Keep it for overnight.
2. To the pre-digested samples add 5 ml of triacid mixture (add 4 ml for each additional g if plant sample taken for digestion is more than 5 g).
3. Digestion of sample is carried out at 180 to 200°C till it gives while fumes and the liquid content is largely volatilized. At the stage of complete digestion the residue remained in the flask become clear and white. If the digested residues brown with carmelized organic matter then again add 5 ml of triacid mixture and digest the material.
4. Transfer the digested material into a 100 ml volumetric flask, dilute it with distilled water upto the mark.
5. Take 10 ml plant extract into a 50 ml volumetric flask. Add 10 ml ammonium molybdate-ammonium vanadate solution and make the volume to 50 ml with distilled water. Yellow colour will develop after 15-20 minutes.

6. From the 50 ppm standard solution of P pipette out 0, 2, 4, 6, 8 and 10 ml and transfer in 50 ml volumetric flasks then add 10 ml of ammonium molybdate ammonium vanadate solution and make the volume. This will contain 0, 2, 4, 6, 8 and 10 ppm P.
7. Measure the colour intensity of standard solutions at a wave length of 470 nm or by using blue filter in a Spectrophotometer. Prepare the standard curve by plotting concentration of P on X-axis and colorimeter readings on Y-axis on a graph paper.
8. Take readings of plant samples in the similar manner that of standard. Workout the ppm content of P with help of standard curve.

Observations and Calculation

Weight of plant sample = X g

Volume of digested extract prepared = 100 ml

Volume of digested extract taken = 10 ml

Volume of final coloured extract prepared = 50 ml

Concentration of P in ppm in plant sample = ppm reading of plant from standard curve ® X Digested volume of plant sample/Weight of plant sample X Final volume prepare/volume of aliquot taken = R x 100/X x 50/10 = R x 500/ X

Per cent P in plant sample = R x 500/X x 1/ 10000 = R / 20 X

Estimation of potassium and sodium in plant (Bhargava and Raghupathi, 1993)

Principle: Potassium content in plant extract is determined by flame photometry i.e. when atoms of potassium are excited in flame emit a flame of specific wave length, the intensity of emission is proportional to the concentration of K which is determined in flame photometer using K filter.

Apparatus: Flame photometer, volumetric flask, pipette, beaker, conical flask 50 ml, hot plate etc.

Reagents

1. **Triacid** as described in P determination : First take 100 ml of HNO_3 add 10 ml of concentrated H_2SO_4, mix it and then add 40 ml of 60% $HClO_4$ i.e. in the ratio of 10:1:4 of HNO_3: H_2SO_4: $HClO_4$.
2. **Potassium standard solution:** Dissolve 1.9103 g of AR grade KCl in distilled water. Transfer it to 1 litre volumentric flask and make the volume. This solution contains 100 ppm of K. To prepare 100 ppm solution take 10 ml of 1000 ppm K solution in a 100 ml volumetric flask and make up its volume.

Procedure

1. Digest the plant sample with triacid as per the method given in P determination in plant sample and make the volume in 100 ml volumetric flask.
2. Prepare 0, 2, 4, 6, 8 and 10 ppm K solution by taking 0, 2, 4, 6, 8 and 10 ml of 100 ppm K solution in 100 ml flask respectively and making up their volume.
3. Feed the standard in flame photometer and take the reading. Plot the concentration on X-axis and flame photometer reading on Y-axis on graph paper and prepare a standard curve.
4. Feed the unknown sample in the flame photometer, note the readings and find out the concentration with the help of standard curve. If the readings of sample go above the standard then dilute the extract 5 to 10 times or more as required.

Observations and Calculation

Weight of plant sample = X g

Volume of extract prepared = 100 ml

Concentration of K in plant sample = ppm K in plant sample from standard curve ® x Volume of extract prepared / Weight of sample

= R x 100 / X

$$\text{Per cent K} = R \times \frac{100}{X} \times \frac{1}{10000} = R \times \frac{1}{100X}$$

Estimation of calcium and magnesium in plant (Bhargava and Raghupathi, 1993)

Principle: Calcium can be determined in plant extract with help of flame photometer as well as Atomic Absorption Spectrophotometer with the help of their standard solutions.

Apparatus: Flame photometer or Atomic Absorption Spectrophotometer, volumetric flask, pipette, beaker, measuring cylinder etc.

Reagents

1. $CaCO_3$, Conc. HCI, Conc. HNO_3, $HCIO_4$, $MgSIO_4.7H_2O$ and magnesium ribbon.
2. **Standard solution of Ca:** Take 20247 g of $CaCO_3$ in a beaker, add 5 ml distilled water and approximately 10 ml HCL. Transfer the contents in 1 litre volumetric flask and make up the volume. This will contain 1000 ppm

of Ca. Working solution of 100 ppm Ca is prepared by taking 10 ml of 1000 ppm of Ca solution in 100 ml volumetric flask and making up the volume with distilled water.

3. **Standard solution of Mg**: Dissolve 10.141 g of $MgSO_4.7H_2O$ in 200 ml distilled water and make up the volume in a 1 litre flask. This will give 1000 ppm Mg solution. Working standard can be prepared as given in above for Ca.

Procedure

1. Dilute the diacid digested plant material in 100 ml volumetric flask 25-50 times depending upon the concentration of Ca or Mg. The diluted plant extract is fed to the atomic absorption spectrophotometer using the hollow cathodes of respective elements.
2. In the similar way readings of Ca or Mg standard are taken by preparing their 0, 5, 10, 15, 20 and 25 ppm solutions and a standard curve is prepared by plotting concentration on X-axis and AAS readings on Y-axis on a graph paper.
3. Concentration of Ca or Mg in plant samples is obtained with the help of standard curve.
4. If AAS facilities is not available then Ca and Ca + Mg in diluted plant extract ban be determined by EDTA titration method as described for Ca and Mg determination in soil.

Observations and Calculation

Weight of plant sample taken = X g

Volume of extract prepared = 100 ml

Calcium/Magnesium percent in plant sample =R X100/Weight of plant sample X100/ 10^6

Where R is the ppm from standard curve In case of dilution the reading should be multiplied by dilution factor 25 to 50.

Estimation of sulphur in plant (Tabatabai and Bramner, 1970)

Principle: The plant material is digested with nitric and perchloric acid and the sulphate content of an aliquot of the digest is determined turbidimetrically as $BaSO_4$ by a barium chloride gelatine procedure. During digestion of plant sample its sulphur is converted into sulphate and when treated with $BaCI_2$ white $BaSO_4$ provides turbidity proportional to the amount of sulphate. Colorimetric measurement of the turbidity gives the quantity of sulphur.

Apparatus: Colorimeter, volumetric flask, pipette, beaker, glass beads, stream plate, hot plate etc.

Reagents

1. Nitric acid – 70 % AR grade:
2. Perchloric acid – 70% AR grade
3. **Hydrochloric acid (6M):** Add 500 ml of conc. HCl (AR grade, sp. Gr. 1.19) to about 300 ml of water and dilute the solution to one litre.
4. **Barium chloride gelatine reagent:** Dissolve 0.6 g of gelatine in 200 ml of hot (60-70°C) water and allow the solution to stand at 4°C. After 16 to 18 hours, bring the semi gelatinous fluid to roop temperature add 2 g of reagent grade BaCl2.2H_2O and swirl the mixture until the $BaCl_2$ is dissolved. Allow the reagent to stand at rop temperature for at least 2 hours before use.Store in a refrigerator.
5. **Standard sulphate solution:** Dissolve 5.43 g of reagent grade potassium sulphate in water and dilute the solution to one littre.1 ml of this solution contains 1 mg of sulphate-S (1000 ppm sulphate-S)
6. **Water:** Use distilled de-ionized or double distilled water in preparation of reagants and in the procedure described.

Procedure

1. Take 0.25 g of plant material in a dry 100 ml volumetric flask containing two glass beads and add 3 ml of nitric acid and heat the flask on a stream plate for 30 minutes.
2. Allow the flask to cool at room temperature for 15 minutes and add 2 ml of perchloric acid.
3. Now heat the flask on a hot plate until white flumes of perchloric acid are visible and then heat continuously for one hour (total digestion time is about 2 hours).
4. Then allow the flask to cool to room temperature and add 3 ml of 6 M HCI and make the digest to 100 ml by adding distilled water.
5. Stopper the flask, invert it several times to mix the contents and allow it to stand over night.
6. Transfer a 10 ml aliquot of clear, supernatant liquid to a 50 ml Erlenmeyer flask, add 10 ml of water, mix and add 1 ml of the barium chloride gelatine reagent and swirl the flask for few seconds to mix the contents.
7. After 40 minutes, transfer the resulting cloudy mixture to a klett Summerson Colorimeter cell (2 cm light path) and measure its turbidity with a klett-

Summerson photo electric colorimeter fitted with a blue (No. 42) filter or at 420 nm wavelength.

8. Calculate the sulphate-S content of the 10 ml aliquot analysed by reference to a calibration graph plotted from the result obtained standards.
9. Simultaneously prepare a series of standards. Take 5 ml of 1000 ppm sulphate-S solution in a 100 ml volumetric flask. This will contain 50 ppm sulphate-S. Now take 0, 0.5, 1.0, 1.5, 2.0, 3.0, 4.0 and 5.0 ml of 50 ppm sulphate-S solution in a 50 ml volumetric flask and add some distilled water and 3 ml of nitric acid. Then proceed as per steps 6 to 7 described above and make the volume 50 ml ml. The concentration of sulphate-S in the series will be 0, 0.5, 1.0, 1.5, 2.0, 3.0, 4.0 and 5.0 ppm, respectively.
10. Prepare a standard curve by plotting concentration of sulphate-S on X-axis and colorimeter readings on Y-axis on graph paper.

Observation and Calculation

1. Weight of plant material = 0.25 g
2. Volume of digested material = 100 ml
3. Volume of aliquot taken = 10 ml
4. Final volume prepared for taking reading = 50 ml
5. Concentration of sulphate-S in ppm = R x 100/0.25 x 50/10 = R x 2000
6. Per cent sulphate-S = R x 2000 x 1/10000 = R/5

   R= Concentration of sulphate-S obtained from standard curve.

Estimation of zinc, copper, manganese and iron in plant (Lindsay and Norwell, 1978)

Apparatus: Atomic Absorption Spectrophotometer, digestion assembly, conical flask, pipette, volumetric flask, sample extract bottles etc.

Reagents

1. **Triacid mixture:** Take HNO_3 : H_2SO_4 : H_2ClO_4 in the ratio 9 : 1 : 4 by mixing HNO_3 first with H_2SO_4 and then $HCLO_4$.
2. **Standard solutions of Micronutrients**:

   Zinc: Dissolve 4.398 g of AR grade $ZnSO_4.7H_2O$ in double distilled water and transfer to a 1 litre volumetric flask, make its volume with the help of double distilled water. This gives 1000 ppm Zn solution. 100 ppm solution is prepared by diluting 10 ml of 1000 ppm solution to 100 ml.

   Copper: Dissolve 3.929 g of AR grade $CuSO_4.5H_2O$ in double distilled water and follow the same procedure as given in Zn.

Mangenese: Dissolve 3.076 g of AR grade $MnSO_4.H_2O$ in double distilled water and follow the same procedure as given in Zn.

Iron: Dissolve 7.022 g of AR grade $(NH_4)_2Fe(SO_4)2.6H_2O$ in 400 ml double distilled water in a beaker and add 5 ml of concentrated H_2SO_4. Transfer it to a 1 lit volumetric flask and make up to the volume. This will give 1000 ppm Fe solution. From this prepare 100 ppm solution as given above for Zn.

Procedure

1. Digest the plant sample with triacid as per the method given in P determination in plant sample and make the volume in 100 ml volumetric flask. Use this extract for the determination of Zn, Cu, Mn and Fe.

   Prepare the final standard of solutions from stock solutions of Zn, Cu, Mn and Fe and note the readings on a graph paper i.e. concentration on X-aixs and AAS readings on Y-axis and prepare standard curve.
2. Feed the plant sample extract to AAS using Zn, Cu, Mn and Fe hollow cathodes one by one for each element and note the readings and find out the concentration with the help of standard curve.

Observations and Calculation

1. Weight of dried plant sample taken = 1 g
2. Volume of extract prepared = 100 ml
3. ppm of Zn or Cu or Mn or Fe in plant = R x 100/1 = R x 100

   Where R = ppm of respective element observed from standard curve.

Estimation of molybdenum in plant (Kotur, 1990)

Principle: Molybdenum is determined colorimetrically by measuring its concentration in Mo complex of thiocyanate as it gives umber colour when extracted into an organic solvent. The colour intensity is proportional to the concentration of Mo in the extract.

Apparatus: Colorimeter or Spectrophotometer, muffle furnance, volumetric flask, pipette, conical flasks, beaker, measuring cylinder, silica crucible, separating funnel and funnel.

Reagents

1. a. Hydrochloric acid b. Sodium hydroxide
2. **Stannous chloride solution:** Dissolve 20 g of $SnCl_2$ in 20 ml concentrated HCl by heating, make its volume to 200 ml with double distilled water. Prepare a fresh solution every day.

3. **Ferric chloride hexahydrate solution:** Dissolve 49 g of $FeCl_3.6H_2O$ in double distilled water and make it one litre.
4. **Sodium nitrate solution**: Dissolve 42.5 g of $NaNO_3$ in double distilled water and make it 100 ml.
5. **Ammonium thiocyanate solution:** Dissolve 50 g NH_4SCN in double distilled water and make it 500 ml.
6. **1, 1- Di-isopropyl ether:** Prepare solution by taking 330 ml of $SnCl_2$ solution, 3.3 ml of NH_2 SCN solution and 340 ml of double distilled water. Take 100 ml of 1, 1- Di-isopropyl ether in a saperating funnel and add 10 ml of $SnCl_2$. NH_4SCN solution. Shake thoroughly and keep it till organic phase separate from aqueous phase then discard the aqueous phase and keep it till organic phase separates from aqueous phase then discard the aqueous phase and then wash the ether with 2 N HCl. Repeat this process four or five times to obtain the desired 1,1-Di-isopropyl ether
7. **Standard solution of Mo:** Dissolve 0.250 g of AR grade molybdenum trioxide (MoO_3) in 10 ml of 0.1 N NaOH, make it slightly acidic by addind HCl andmake it 100 ppm solution of Mo.

Procedure

1. Transfer 50 ml of plant extract in a separating funnel. Add in sequence 1 ml of $NaNO_3$, 5 ml of NH_4 SCN and 5 ml of $SnCl_2$ solutions. After addition of each solution shake the contents of separating funnel. Remove the stopper and wash the neck of funnel and stopper with double distilled water in the separating funnel with the help of a fine tipped wash bottle for keeping the neck and stopper free of reagents. Further the final pouring of Mo complex in ether phase will be done through neck so it should be free reagents and check transfer of residual reagents. Finally add 5 ml of washed 1, 1-Di-isoprophyl ether and shake the contents of separating funnel.
2. Allow the ether phase to separate, remove the stopper, rinse and wipe out the neck with filter paper and draw out the aqueous phase. Transfer the ether phase into a 15 ml centrifuge tube by pouring from top of the separating funnel. Measure the transmittance at 475 nm.
3. For standard curve transfer 0, 0.5, 1.0, 2.0, 4.0 and 5.0 ml of standard solution in separating funnel and add 5 ml of 6 N HCl and then add 1 ml of $FeCl_3$, 1 ml of $NaNO_3$, 5 ml of NH_4SCN and 5 ml of $SnCl_2$ and proceed in the similar manner given for plant extract. Plot the transmittance reading on Y-axis and concentration of Mo on X-aixs on a graph paper and prepare a standard curve.

Observations and Calculation

1. Weight of plant sample = 1g
2. Volume of extract = 50 ml
3. Mo (ppm) in plant extract = R x 50/1=R x 50
   Where R is colorimetric reading of ppm Mo in plant sample.

Estimation of boron in plant (Hatcher and Wicox, 1950)

Principle: Colorimeter boron estimation in plant extract is based upon the formation of alcohol soluble coloured product obtained when a solution prepared by mixing boric acid, oxalic acid and curcumin which is further evaporated to dryness. The intensity of colour developed will be proportional to the concentration of boron.

Apparatus: Colorimeter or Spectrophotometer, water bath, muffle furnace, pipette, conical flask,beaker,volumetric flask, measuring cylinder, porcelain dish, silica crucible & funnel etc.

Reagents

1. **Curcumin: Oxalic acid solution:** 0.04 g of finely ground curcumin and 5 g of oxalic acid are dissolved in 100 ml of 95% ethyl alcohol. Transfer the solution in a coloured bottle and store in refrigerator. This can be used for a week only.
2. Ethyl alcohol 95%
3. **Standard solution of Boron:** Dissolved 0.286 g of boric acid (H_3BO_3) AR grade in double distilled water and transfer the solution in to a 1 littre flask and make up the volume. This solution contains 50 ppm of B.

Procedure

1. Take 1g plant material, transfer it into a silica crucible and make its ash by keeping it in a muffle furnace at 550°C for 5 hours. Cool the crucible in a desiccators. Cooled ash is dissolved in 5 ml of concentrated HCl, if residue is not dissolve, warm the solution. Filter the solution through Whatman No.1 and give washings of double distilled water. Transfer the filtrate into 50 ml flask and make its volume.
2. Take 50 ml plant extract into a porcelain dish and add 4 ml of curcumin oxalic acid solution.
3. Mix the solution well with the help of a glass rod and place the porcelain dish on water bath maintained at 55°C & when whole solution of the dish evaporated. Keep this dish on water bath for another 15 minutes.

4. Cool the dishes and then add 25 ml of 95% ethyl alcohol, rub the sides of the dishes with a policeman and dissolve all the residues in alcohol. Filter the contents of dishes in a small beaker through What man No. 1 filter paper.
5. Take the readings of the filtrate at 540 nm.
6. To prepare standards of 0, 0.1, 0.25, 0.5, 1.0 and 2.0 ppm of solution pipette out 0, 0.2, 0.5, 1.0, 2.0 and 4.0 ml of 50 ppm of B solutions in 100 ml volumetric flask and make up the volume.
7. Pipette out 1 ml from each flask containing 0, 0.1, 0.25, 0.5, 1.0 and 2.0 ppm of B in separate porcelain dishes and repeat the steps 1 to 4 carried out for plant sample.
8. Note the readings of standard on Y-axis and concentration on X-axis on a graph paper and prepare a standard curve.
9. Workout the ppm content of B of plant sample with the help of standard curve.

Observation and Calculation

Weight of plant sample = 1 g

Plant extract prepared = 50 ml

ppm of Boron content in plant sample = R x 50/1 = R x 50

Wherer R is the ppm of B obtained from standard curve.

Boron can also be determined from dry ash extract by AAS using Boron hollow cathode.

Estimation of sodium in plant (Bhargava and Raghupathi, 1993)

The procedure for sodium is similar to that of potassium. However, a different filter of Na has to be used as the radiation emitted by excitation of Na atoms of a different wavelength.

Apparatus: Flame photometer, volulmetric flasks, pipettes, beakers, hot plate etc.

Reagents

1. Diacid (HNO-$_3$ and $HClO_4$).
2. **Sodium standard solution:** Dissolve 2.541 g of NaCl in 1000 ml deionized or double distilled water to get 1000 ppm Na. 10 ml of this stock solution is further distilled to 100 ml to get 100 ppm Na.

Procedure

1. Take one g of ground plant material in 100 ml volumetric flask and add 10 ml of diacid mixture of HNO_3 and $HClO_4$ in 9:4 ratios. Mix the contents by swirling the flask.
2. Place the flask on heating device in a digestion chamber.
3. The flask is heated first at low and then at higher temperature until the production of red NO_2 fumes ceases.
4. The contents are further evaporated until the volume is reduced to about 3 to 5 ml but not to dryness. The completion of digestion is confirmed when the liquid becomes colour less.
5. Cool the flask adds deionized or double distilled water and makes the volume up to mark.
6. Filter the solution through what man No. 1 filter paper.
7. Now prepare a series of standard of 0, 2, 4, 6, 8 and 10 ppm Na by taking 0, 2, 4, 6, 8 and 10 ml of 100 ppm Na solution in 100 ml flask, respectively and make up their volume.
8. Feed the standard in flame photometer and take readings. Plot the concentration on X-axis and flame photometer readings on Y-aixs on graph paper and prepare the standard curve.
9. Feed the sample extract in the flame photometer. Note the reading and find out the concentration of Na with the help of standard curve. It the reading of sample is above the standard then dilutes the extract.

Observations and Calculation

Weight of plant sample = 1 g

Volume of extract prepared = 100 ml

Concentration of Na (ppm) in plant sample = ppm Na in plant sample from standard curve

(R) X Volume of extract prepared /Weight of sample = R x 100/1 = R x 100

Concentration of Na in per cent = R x 100 x 1/10000 = R x 1/100

4

Water Quality Analysis

Irrigation water always contains some soluble salts irrespective of its source. The suitability of waters for a specific purpose depends on the types and amounts of dissolved salts. Some of the dissolved salts or other constituents may be useful for crops. However, the quality or suitability of waters for irrigation purposes is assessed in terms of the presence of undesirable constituents, and only in limited situations is irrigation water assessed as a source of plant nutrients. Some of the dissolved ions, such as NO_3^- are useful for crops.

Collection of water samples

A representative water sample (500 ml) is collected in a glass or polyethylene bottle, which should be properly washed/rinsed with the same water that is being sampled. Floating debris or any other contaminant should be avoided while collecting the sample. After proper labelling (e.g. source of water, date of collection, and type of analysis required), the sample should be sent to the laboratory without undue delay.

Some of the anions such as SO_4 and NO_3 may be quite low in irrigation waters. Hence, the large volume of the sample has to be first concentrated by evaporating it to about 100 ml in order to obtain their detectable amounts.

Analytical methods

pH

The pH is determined by putting about 50 ml of the water sample in a clean 100 ml beaker and using a pH meter as described in estimation of soil pH.

Electrical conductivity

A conductivity meter cell is filled with the water sample, and the EC is determined as described in in estimation of soil EC.

Calcium and magnesium

The usual method for determination of Ca and Mg is by versenate (EDTA) titration (Cheng and Bray, 1951). The estimation of Ca and Mg can also be done using an AAS, as described in soil analysis. Estimation of Ca and Mg in water by the EDTA method is described below.

Apparatus: a porcelain dish; some volumetric flasks; a burette.

Reagents

Standard versenate solution (EDTA): Prepare an approximately 0.01N solution of EDTA disodium salt (versenate) by dissolving 2.0 g in distilled water, add 0.05 g of magnesium chloride ($MgCl_2.6H_2O$) and dilute to 1 litre. Standardize against 0.01N calcium chloride solution prepared by weighing 0.500 g AR-grade $CaCO_3$ (oven-dried) and dissolving in minimum excess of dilute HCl (AR), followed by making the volume up to 1 litre with distilled water.

Ammonium chloride-ammonium hydroxide buffer (pH 10): 67.5 g of pure ammonium chloride dissolved in 570 ml of concentrated ammonium hydroxide and made up to 1 litre; pH adjusted to 10.

EBT indicator: 0.5 g of EBT and 4.5 g of hydroxylamine hydrochloride (AR) dissolved in 100 ml of 95 percent ethyl alcohol.

Procedure

- Put 5 ml of the water sample in a porcelain dish (8 cm in diameter).
- Dilute to about 25 ml with distilled water.
- Add 1 ml of ammonium chloride-hydroxide buffer and 3–4 drops of EBT indicator.
- Titrate with the standard versenate solution. The colour change is from wine-red to bright blue or bluish green. At the end point, no tinge of the red colour should remain.
- From the volume of 0.01N EDTA (standardized against 0.01N $CaCl_2$) solution required for titration, the concentration of Ca and Mg is obtained directly as follows:

Observations and Calculations

$$\text{Ca} + \text{Mg}(\text{me / litre}) = \frac{\text{ml versenate (EDTA) used} \times \text{normality of EDTA} \times 1000 \text{ ml}}{\text{a liquot taken (ML)}}$$

$$\text{Ca} + \text{Mg (g/litre)} = \frac{\text{Ca} + \text{Mg in me/litre} \times \text{equivalent wt.}}{1000}$$

$$= \frac{\text{Ca} + \text{Mg in me/litre} \times 32.196}{1000}$$

OR

$$\text{Ca} + \text{Mg}(\text{g/litre}) = \frac{\text{ml versenate (EDTA) used} \times \text{normality of EDTA} \times 1000 \text{ x equivalent wt.}}{\text{ml aliquot taken} \times 1000}$$

$$= \frac{\text{ml versenate(EDTA) used} \times \text{normality of EDTA} \times 32.196}{\text{ml aliquot taken}}$$

Sodium

A small amount of Na is generally present even in the best-quality irrigation water. The concentration of Na may be quite high in saline water with an EC higher than 1 mS/cm and containing relatively less Ca and Mg. Its estimation is of interest when the water sample tests saline (i.e. an EC higher than 1.0 mS/cm at 25 °C). The determination of Na is carried out directly with the help of a flame photometer using appropriate filters and standard curves prepared by taking a known concentration of Na.

Apparatus: a flamephotometer; some volumetric flasks; some beakers.

Reagent

Reagent required is NaCl (AR-grade).

Procedure

Preparation of the standard curve:

1. Take 2.5413 g of NaCl (AR), dissolve in water to make to the volume up to 1 litre; this gives a solution of 1 000 μg Na/ml. From this solution, take 100 ml and dilute to 1 litre to obtain 100 μg Na/ml as stock solution.
2. To prepare working standards, take 5, 10, 15 and 20 ml of stock solution in 100-ml volumetric flasks and make up the volume. This gives 5, 10, 15 and 20 μg Na/ml.
3. Feed the standards on the flamephotometer one by one in order to obtain a standard curve, with absorbance on the y-axis and the respective concentrations of Na on the x-axis.
4. Feed the water samples on the flamephotometer, and record the absorbance for each sample.

Observe the concentration of Na against each absorbance (in μg Na/ml).

The relevant calculation is:

$$\text{Content of Na in mg/litre of water} = \frac{A \times 1000}{1000} = A$$

Where

1. A = absorbance reading (μg/ml) from the standard curve. It is important to note that:
2. If a water sample is diluted for estimation, the quantity of Na as observed on the standard curve is multiplied by the dilution factor.
3. If the water sample is concentrated before estimation, the quantity noted from the standard curve is divided by the concentration factor.
4. Normally, no dilution or concentrated is required.

Carbonates and bicarbonates

The estimation is based on simple acidimetric titration (Richards, 1954) using different indicators that work in the alkaline (higher than 8.2) or acidic pH range (lower than 6.0).

Apparatus: a porcelain dish; a burette.

Reagents

Phenolphthalein indicator: 0.25 percent solution in 60 percent ethyl alcohol.

Methyl orange indicator: 0.5 percent solution in 95 percent alcohol. Standard **sulphuric acid (0.01M).**

Procedure

- Put 5 ml of the water sample (containing not more than one milli-equivalent of carbonate plus bicarbonate) in a porcelain dish.
- Dilute with distilled water to about 25 ml.
- A pink colour produced with 2–3 drops of phenolphthalein indicates the presence of carbonate, and it is titrated with 0.01M sulphuric acid until the colour just disappears (phenolphthalein end point) because of alkali carbonate having been converted to bicarbonate. This is called the half-neutralization stage. This burette reading (volume used) is designated Y.
- To the colourless solution from this titration (or to the original sample of water if there was no colour with phenolphthalein), add 1–2 drops of methyl orange indicator, and continue titration with brisk stirring to the methyl orange end point (yellow); the final reading (volume used) is designated Z.

The calculation is:

$$\text{Carbonates}\ (\text{me/litre}) = 2(\text{Volume of } H_2SO_4) \times \text{Molarity of } H_2SO_4 \times \frac{1000}{\text{ml of aliquot}}$$

$$2Y \times 0.01 \times \frac{1000}{5}$$

$$2Y \times 2 = 4Y$$

$$\text{Carbonates } (g/\text{litre}) = \frac{2\ (\text{Vol. of } H_2SO_4) \times \text{Molarity} \times 1000 \times \text{Eq.wt.of } CO_3\ (30)}{\text{ml of sample X1000}}$$

$$\frac{2Y \times 0.01 \times 30}{5} = 0.12\,Y$$

The volume of acid used for half-neutralization of carbonate is *Y*. Hence, for full neutralization it has been assumed as 2*Y*:

$$\text{Bicarbonates (me/litre)} = (Z - 2Y) \times \text{molarity of } H_2SO_4 \times \frac{1000}{\text{ml of aliquot}}$$

$$\frac{(Z - 2Y) \times 0.01 \times 1\,000}{5}$$

$$(Z - 2Y) \times 2$$

Where carbonate is absent: $Z \times 2$.

Residual sodium carbonate

This is an important characteristic for assessing the suitability of irrigation water in consideration of a probable sodium hazard. It is calculated from the analysis data for carbonates, bicarbonates and Ca plus Mg as follows (all expressed in me/litre):

RSC (me/litre)= $(CO_3^{2-} + HCO_3^{-}) + (Ca^{2+} + Mg^{2+})$

Boron

The method for B estimation is same as described in estimation of Boron in soil. The determination is carried out using the azomethine-H colorimetric method. It can also be estimated on an AAS. Suitable quantities of the sample should be taken depending on the B content in the waters.

Chloride

Mohr's titration method is the one most commonly used for chloride estimation. It depends on the formation of a sparingly soluble brick-red silver chromate ($AgCrO_4$) precipitate at the end point when the sample is titrated against standard silver nitrate ($AgNO_3$) solution in the presence of potassium chromate (K_2CrO_4) as indicator. Initially, the Cl ions are precipitated as AgCl and the dark brick-red precipitate of Ag_2CrO_4 starts just after the precipitation of AgCl has finished.

Apparatus: beakers; a porcelain dish; a burette.

Reagents

1. Potassium chromate (K_2CrO_4) indicator (5 percent) solution: Dissolve 5 g of K_2CrO_4 in about 75 ml of distilled water, and add saturated solution of

$AgNO_3$ drop by drop until a slight permanent red precipitate is formed. Filter and dilute to 100 ml. With high-purity AR, the indicator solution can be prepared directly.

2. Standard silver nitrate solution (0.05M): Dissolve 8.494 g of silver nitrate ($AgNO_3$) in distilled water, and make the volume up to one litre. Standardize it against standard NaCl solution, and keep it in an amber-coloured bottle away from the light.

Procedure

1. Put 5 ml of the sample in a 100 ml beaker or a porcelain dish, and dilute to about 25 ml with distilled water.
2. Add 5–6 drops of K_2CrO_4 indicator (making it dark yellow), and titrate against the standard $AgNO_3$ solution with continuous stirring until the first brick-red tinge appears.
3. Run a blank to avoid error caused by any impurity in the chemicals.

The relevant calculation is:

$$\text{Cl mg/litre of water} = \frac{X \times 1.775 \times 1000}{\text{ml of sample}}$$

Where:

ml of water sample taken = 5;

X = ml of 0.05M $AgNO_3$ consumed in titration;

1.775 = factor representing mg of Cl in aliquot/sample as calculated below:

1 ml of 1M $AgNO_3$ = 1 me of Cl;

1 ml of 0.05M $AgNO_3$ = 0.05 me of Cl = 35.5 × 0.05 = 1.775 mg of Cl (in aliquot).

Sulphate

While traces of sulphate occur universally in all types of waters, its content may be appreciably high in several saline waters showing an EC higher than 1 dS/m at 25°C. Sulphate can be determined gravimetrically, colorimetrically, turbidimetrically or titrimetrically. Here, the turbidimetric method is described.

Sulphate content is determined by the extent of turbidity created by precipitated colloidal barium sulphate suspension. Barium chloride solid crystals are added to ensure fine and stable suspension of $BaSO_4$ at a pH of about 4.8. This also eliminates interference from phosphate and silicate. This fine suspension of $BaSO_4$ is stabilized by gum acacia, and the degree of turbidity is measured with a turbidity meter or estimated spectrophotometrically at 440 nm.

Apparatus: A spectrophotometer; some beakers; some volumetric flasks.

Reagents

- Sodium acetate – acetic acid buffer: Dissolve 100 g of pure sodium acetate in 200 ml of distilled water. Add 31 ml of glacial acetic acid, and make the volume up to 1 litre. Adjust the pH to 4.8.
- Gum acacia: Dissolve 2.5 g of gum acacia in 1 litre of distilled water. Keep overnight and filter.
- Barium chloride: Grind pure $BaCl_2$ crystals to pass through a 0.5-mm sieve but be retained on a 0.25 mm sieve.
- Potassium sulphate solution: To make a stock solution of 10 me S/litre, weigh 1.74 g of pure K_2SO_4 salt and dissolve in 1 litre water.

Procedure

- Put 5 ml of the water sample (of < 1 me S/litre) in a 25 ml volumetric flask. If the EC of water is > 1 dS/m, dilute it with distilled water to bring the EC to lower than 1 dS/m.
- Add 10 ml of sodium acetate – acetic acid buffer to maintain the pH at about 4.8.
- Add 1 ml of gum acacia and 1 g of $BaCl_2$ crystals, and shake well.
- Make the volume up to 25 ml with distilled water.
- Invert the flask several times, and measure the turbidity with a spectrophotometer at 440 nm using a blue filter.
- Preparation of the standard curve: For 0, 1, 2, 3, 4 and 5 me S/litre, pipette 2.5, 5, 7.5, 10 and 12.5 ml from stock solution containing 10 me S/litre into 25-ml volumetric flasks. Then develop the turbidity and measure its intensity as in the case of samples. Draw a curve showing S concentration on the x-axis and absorbance on the y-axis.

The S content of the samples is calculated using the standard curve, taking into account the dilution factor of five (5 ml made to 25 ml) and expressed as me S/ litre of water.

Nitrate nitrogen

This method depends on the reduction of nitrate to ammonia by adding Devarda's alloy and alkali. The nitrites (NO_2^-) (if present) in the sample are also reduced and determined along with NO_3^--N.

The apparatus required consists of:

a Kjeldahl distillation assembly; an electric muffle furnace; a desiccator.

Reagents

- Magnesium oxide (MgO): Heat the MgO at 65 °C for 2 hours in an electric muffle furnace to remove any traces of $MgCO_3$ that may be present. Cool in a desiccator over solid KCl, and store in a tightly stoppered bottle.
- Boric acid with mixed indicator: Weigh 20 g of boric acid, and add about 900 ml of hot distilled water. Cool and add 20 ml of mixed indicator, and make the volume up to 1 litre.
- Mixed indicator: Dissolve 0.066 g of methyl red and 0.099 g of bromocresol green in 100 ml of alcohol.
- Standard sulphuric acid (0.02M).
- Devarda's alloy: Mix Cu, Al and Zn in the ratio of 50:45:5, and grind to pass through a 0.15 mm sieve.

Procedure

- Put 50 ml of the water sample in the distillation flask.
- Add 0.5 g of MgO and 0.2 g of Devarda's alloy.
- Put the heaters on and collect the NH_4 (NO_3 converted into NH_4 by reducing agent – Devarda's alloy) in boric acid (20 ml) having mixed indicator into a conical flask, which is connected with the distillation apparatus.
- Continue distillation to collect about 35–40 ml.
- Remove the distillate first and then switch off the heating system.
- Titrate the distillate against 0.02M H_2SO_4 until the pink colour appears.
- Carry out a blank simultaneously.

The relevant calculation is:

$$NO_3^- - N\ (\text{mg/litre}) = \frac{(X - Y) \times 0.56}{50\ (\text{ml of sample})} \times 1000$$

Where,

X = volume (ml) of 0.02M H_2SO_4 consumed in sample titration;

Y = volume (ml) of 0.02M H_2SO_4 consumed in blank titration;

0.56 = factor. 1 litre 1M H_2SO_4 = 28 g N; therefore:

$$1\ \text{ml}\ 0.02\text{M}\ H_2SO_4 = \frac{28 \times 0.02 \times 1000}{1000}\ \text{mg N} = 0.56\ \text{mg N}$$

5

Basic Philosophy of Nutrient Management

The basic philosophy of nutrient management should be to apply fertilizers at rates to ensure high fertilizer use efficiency so that the amount of unutilized fertilizers is reduced to environmentally acceptable levels. Leaching of nitrate into public drinking water supplies has become a national problem. There is no realistic alternative but to use increasing quantities of inorganic fertilizers for maintaining food supplies in order to feed the increasing population. Used properly, fertilizers can, in fact, lead to better environment because greater vegetative growth and higher crop yields leave large quantities of residues in the soil, thus reducing erosion; absorb more carbon dioxide, a major source of green house effect and lessen the need to bring marginal and forest land into cultivation. The declining or stagnating yield trend in India has been attributed to multiple nutrient deficiencies and imbalance of nutrients. Therefore, the most widely recognized alternative is integrated nutrient management (Trehan *et al.,* 2008).

Soil test based fertilizers is the maintenance of soil fertility and plant nutrient supply to an optimum level for sustaining the desired productivity through optimization of the benefits from fertilizers, organic manures, green manures, bio-fertilizers, non-conventional sources and crop residues. Soil test based fertilizers aims at maximization of the use efficiency and minimization of the avoidable losses of nutrients from all the sources such that triple objective of maximization of crop yields, sustenance of soil, water and air quality and improvement of socio-economic conditions of farming community is accomplished.

Following information on nutrient deficiency symptoms; and on suitable source, dose, time and method of fertilizer/manure application can be of help to efficiently manage nutrient requirement of vegetables crop.

Major nutrients

Nitrogen: Nitrogen is a first limiting nutrient in vegetable crops production thus has a great influence on growth attributes, yield attributes, yield and its quality.

A mature crop of vegetable yielding 20-35 t/ha vegetable crops removes 100-150 kg N/ha. The Indian soils are generally deficient in organic matter thus unable to release N at a rate required to maintain adequate N supply to the growing plant. Therefore, application of nitrogen in form of fertilizers and manures becomes indispensable to meet N needs of the crop.

Nitrogen deficiency symptoms: The plants with sufficient N are characterized by vigorous growth, (No. of leaf and branches) increased leaf area index. Whereas, its deficiency leads to pale green coloration of leaf margins at initial stage turning to pale yellow foliage in case of acute deficiency. These symptoms first appear at lower leaves due to mobile in plant and later on whole plant gives yellowish look leading to dropping of lower leaves thereby decreasing photosynthetic activity. On the other hand, excess of N delays vegetable, onset of linear phase of growth ultimately resulting in lower yield.

Source of N fertilizer: The performance of different sources of N varies with the agro-climatic zone and soil type. Among the number of N fertilizers available, ammonium sulphate (21% N) has been found to be the best nitrogenous carrier. The better performance of ammonium sulphate might be attributed to the S supplied by it as it contains 25 per cent S. However, because of its high cost it may not be within reach of small and marginal farmers but can be of great use in S deficient soils. Moreover in long use, it causes acidity in soil. The calcium ammonium nitrate (25% N) and urea are widely used in crop production. Urea alone accounts for nearly 78 per cent of total N fertilizers produced in the country. It is cheap and concentrated N source having 46% N but in potato crop it is 10-15 per cent less efficient compared to other N fertilizers *i.e.* ammonium sulphate and calcium ammonium nitrate. Among other N sources, ammonium chloride, potassium nitrate, di ammonium phosphate and nitro phosphate are available but their use is area specific.

The poor performance of urea has been found to be positively related to its dose because of its adverse effect on plant emergence particularly at higher doses. Moreover the adverse affect is observed more in alluvial and calcareous soils having high pH as compared to acidic soils with low pH. Similarly, the efficiency of urea has been reported to be at par with calcium ammonium nitrate in years of normal rainfall but in drought years, it proved to be inferior. Likewise, ammonium chloride has been found to be more efficient at lower levels but less at higher levels because of chloride/nitrate interaction.

Time of N application: Application of N at a proper time is imperative and vital in getting full benefit of applied N. In potato crop, sufficient N is needed in early stages to build up crop canopy and to enhance leaf area thus giving longer period for development during crop growth period.Hence the time of application of N should have to be scheduled in such a way that sufficient N is made available at the aforesaid critical period.

Split application of N has proved to be more effective than its full application at sowing / planting. This not only reduces leaching losses of N and better utilization of applied N but at the same time also improves N use efficiency. In case of urea, this is more advantageous in order to overcome its deleterious effect on crop emergence N from urea at a dose more than 40-50 kg N/ha at sowing / planting. In general, application of N in two or three split dose produces higher yields and results in higher N recovery than applying entire dose at planting/ sowing. three split of N is better than its two splits. From above it can be concluded that split application of N *i.e.* half through calcium ammonium nitrate + urea in 1:1 ratio at planting/sowing and rest through urea at 40-60 days after sowing is ideal for higher production in potato (Trehan *et al.*, 2008).

Method of N application: The uptake and efficacy of N fertilizer depend upon the method of application. The broadcasting of N fertilizer helps in quick emergence than its placement 1-2.5 cm below the seed tubers. The broadcasting consists of application of N fertilizer on surface soil followed by ridging, thus fertilizer which is on surface is brought to ridges. This is true for calcium ammonium nitrate or ammonium sulphate but same technique does not hold good for urea. Studies indicate that application of urea by broadcast and mixed with moist soil before 2-3 days of planting is quite effective in soils where crop is taken under assured irrigation.

In case of sparse root system, broadcasting does not ensure proper utilization of N under moisture stress particularly in rainfed area. Therefore it is important to place the fertilizer where interception by active roots is more. The placement of N fertilizer *i.e.* urea in side bands (10 cm deep and 5 cm away from seed tubers) at planting has been found to be beneficial to broadcasting and furrow placement with respect to yield and nutrient recovery. The application of high dose of N at the time of planting in furrows should be avoided as study has shown that it affects the plant emergence by increasing the osmotic pressure. Its effect is more when urea is used as an N source as it has a burning effect on sprouts. To overcome this, split application of N or the foliar application of urea tended to confer some advantage over its basal application in full dose.

Foliar application of urea: Among the nutrient elements, the N as NH_4^+ or NO_3^- ions form can easily be adsorbed, translocated and metabolized in plant system thus can be used for foliar application. Urea has this unique property thus has been found to be quite effective in raising yield potential of crop when used for foliar application. Since urea contains biuret which is toxic to leaf as it causes leaf burn so its concentration should not exceed 0.5 per cent in urea.

Urea is easily compatible with fungicides/pesticides for foliar application so it can be effectively used in crop.

Slow release nitrogenous fertilizers: As stated above, copious supply of N is needed throughout the growing phase by crop. This is only possible by using combination of slow and quick acting nitrogenous fertilizers. Urea is hydrolyzed to NH_4^+ or NO_3^- forms with in 48 hours of its application under optimum moisture thereby leading to leaching losses or toxicity which could results in delayed plant emergence. Thus attempts have been made to control the release of N from urea as per needs of the crop through use of biological or chemical inhibitors.

Biological inhibitors: Urease an extra-cellular enzyme hydrolyses the urea into CO_2 and NH_3 with in one to seven days depending upon the urease activity and moisture content in the soil. The use of nitrification inhibitors compounds like neem cake with urea in ratio 1:3 on N basis has been found to enhance urea efficiency and economize N dose upto 25 per cent. Similarly, application of urea with nitrapyrin (N-serve) @ 3 *l*/80 kg N/ha retard the nitrification of urea by upto 65 per cent compared to neem coated urea and brought the efficiency of urea at par with ammonium sulphate beside raising the N recovery from upto 31 per cent.

Chemical inhibitors: The another approach is to coat the urea granules with water resistant chemical in order to slow down its hydrolysis rate hence the release of N. Studies on alluvial soils revealed that only 20-25 per cent of urea was solubilised in the first week after coating of urea with lac, sulphur powder or mobil oil. In acidic soils of Shillong where the rain intensity is more, gypsum or rock phosphate coated urea significantly increased yield. Application of pyrites as amendment also improves the efficacy of urea in alkaline alluvial soils by checking the adverse effect of urea on sprout damage. This has been attributed to the reduction of ammonia toxicity by its neutralization through sulphuric acid produced from pyrites oxidation during its transformation. In acidic soils, the use of magnesium carbonate has also been reported to improve the efficiency of urea.

Residual effect of nitrogenous fertilizers: The recovery of N by the crop depends upon various factors *i.e.* soil type, amount of water received through rainfall in hills and irrigation in plains, temperature, C/N ratio and most important soil organic matter content. In crop, the recovery of N varies between 30-50 per cent thus considerable part of applied N remains unutilized by the plant. This unused N is leached down to lower layers in soil or / and fixed in clay. Thus applied N is expected to leave some residual effect depending upon the N dose. Studies showed that the accumulation of N (Total, $K_2Cr_2O_7$-N and $KMnO_4$-N) is more in presence of farmyard manure followed by green manuring of *dhaincha* and moog. Similarly, the accumulation of N is higher with farmyard manure than calcium ammonium nitrate. Organic manures application, although, does not contribute much in meeting N needs but they increase the apparent N recovery considerably from applied fertilizers.

Long term studies in alluvial soils of Indo-Gangetic plains show that when potato crop is fertilized with recommended dose of N then succeeding wheat requires only 50 per cent of its recommended dose of N while sunflower does not need any N. Similarly, onion requires only 60 kg N/ha if preceding potato crop is given 180 kg N/ha. The low dose required for onion or wheat is mainly because of improvement in soil N status by N applied to potato. It has been established that application of 120 kg N to maize (*Kharif*), 180 kg N to potato and 40 kg N/ha only to wheat is sufficient in potato-wheat-maize sequence. Thus fertilizer N applied to potato leaves sufficient residual effect on succeeding crop but potato itself does not benefit from the residual N from previous crop emphasizing that it require full N dose as per recommendation. The residual effect of N on wheat and other crops may also be linked to the deep rooted crops like wheat which can utilize N more effectively from lower layers after the shallow rooted crop potato is taken.

Phosphorus and potassium: Phosphorus and potassium are the second and third limiting nutrient, respectively, in vegetable production. Phosphorus is involved in wide range of plant processes from permitting cell division to the development of a good root system as well as root growth and hastens crop maturity. On the other hand, application of K increases plant height, crop vigor and impart resistance against drought, frost and diseases. Its application activates number of enzymes involved in photosynthesis, carbohydrate metabolism and proteins and assists in the translocation of carbohydrates from leaves to vegetable part. It increases the yield.

Phosphorus and potassium needs of the crop vary with the agro-climatic region, variety, crop sequence taken and soil type. A healthy crop of vegetable removes about 20-35 kg P_2O_5 and 100-250 kg K_2O/ha indicating that vegetable crops needs for P and K are much higher than that of cereals. The fertilizer use efficiency for P is 10-20% while that for K it ranges between 50-80% thus the residual effect of P and K applied to vegetables effects the fertilizer needs of succeeding crops (Table 2).

Table 2: Nutrient use efficiency of macro and micronutrients

Nutrient	Efficiency (%)	Cause of low efficiency
Nitrogen	30-50	Immobilization, volatilization, denitrification, Leaching
Phosphorus	15-20	Fixation in soils Al – P, Fe – P, Ca – P
Potassium	50-80	Fixation in clay - lattices
Sulphur	8-10	Immobilization, Leaching with water
Micro nutrients (Zn, Fe, Cu, Mn, B)	1-2	Fixation in soils

Phosphorus deficiency symptoms: Its deficiency is more pronounced in acidic soils and red laterite and Ca &Mg rich than in alluvial soils. The former two soils are characterized by high P fixing capacity because of immobilization of P by aluminum and iron compounds in the soils. In the event of its deficiency, the growth of plant is retarded, particularly during early growth stages. The plants are stunted with dull dark green leaves without luster. The leaves are curled with purple pigmentation sometime with marginal scorching. The vegetative growth continues beyond normal time of maturity is delayed.

Potassium deficiency symptoms: The deficiency of K normally coincides with onset of fruiting initiation period. The mild K deficiency starts with dark bluish green leaves and shortened internodes. Terminal leaves show bronzing accompanied with necrotic spots with increasing K deficiency. In case of acute deficiency, leaf margins may dry up and often premature death of plant occurs.

P and K needs of Vegetable crops: vegetable crops requirement for P and K largely depends upon (i) soil pH (ii) soil capacity to supply these nutrients in water soluble forms (iii) parent minerals present in the soil and (iv) climatic factors.

a) **Phosphorus:** The roots absorb P in the form of H_2PO_4 ions in acidic conditions and HPO_4 ions under alkaline conditions. Thus soil pH has a great influence on their availability to the plant. However, immobilization of P or P fixation takes place in presence of Al and Fe compounds in acidic soils and Ca and Mg compounds in alkaline soils which affects the P availability to the plant, efficacy of P fertilizers and P use efficiency by the vegetable crops. Soil pH has a great influence on P availability to the plant.

b) **Potassium:** Indian soils are generally high in total K but fraction of it is in available form because of dynamic equilibrium between fixed, non-exchangeable and exchangeable K. With high crop intensity and high K removal, the soils are likely to become deficient in K with time. The form of K mineral in soil also affects the K needs of vegetable crops. Potassium associated with soil illite is an important source of K to the crops. Unlike P which is immobile, K is highly mobile and tendency of K leaching to lower soil horizons is high particularly when vegetable crop is taken under assured irrigation. Indian soils are poor in N and P with 89 and 80 percent soil samples in low to medium category; relatively better in K with 50 percent samples only low to medium (Fig. 6).

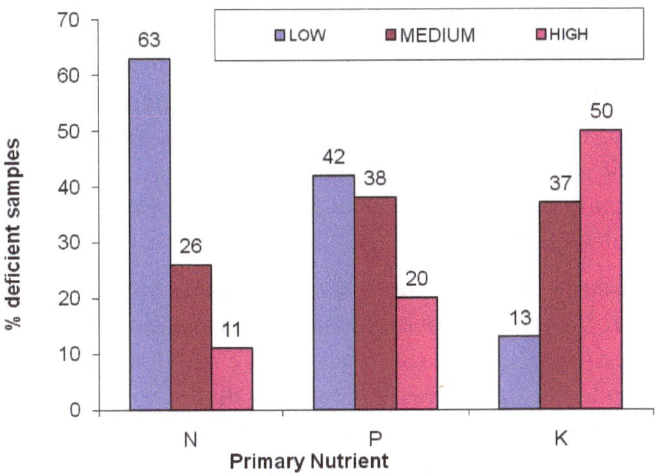

Fig. 6 : NPK status of Indian soils

In acid and laterite soils because of high P fixation problem, transformation of applied P to insoluble phosphate compounds of aluminum, iron and calcium in addition to organic bound P restricts availability of P to potato crop and only a fraction of P remains in soil solution. The water-soluble P carriers are thus more suitable than the citrate soluble in these soils. The most common P fertilizer used in potato cultivation is a single super phosphate having 16% water soluble P_2O_5. The coating of super phosphate balls with fresh cow dung has also been found to be quite effective in these soils. The other P carriers tried are di ammonium phosphate, pyrophosphate and rock phosphate. The later can be used in acid soils by mixing it with the single super phosphate in ratio 3:1. Rock phosphate can however, be used as supplement to buildup the available P status of the soils.

Several potassium fertilizers have been tried in crop but among these potassium chloride (60% K_2O) and potassium sulphate (50% K_2O) have been studied in depth in various agro climatic regions. Although, potassium sulphate (SOP) has been found to be best in term of its beneficial effect on tuber quality *viz.*, dry matter, ascorbic acid and sugar content but due to its high cost, it has not found much use in potato crop. However, in sulphur deficient soils, it can be more effective because of sulphur present in it. The potassium chloride (MOP) is commonly used in potato crop and constitutes 97% of potassium fertilizers consumption in potato crop. Another potassium fertilizer *i.e.* potassium schoenite an indigenous source having double salt of potassium and magnesium sulphate has also been found to be equally good for potato in acidic and alluvial soils.

Time and method of P and K application: The proper placement of the fertilizer has a great influence on the P and K utilization by the vegetable roots and subsequently their effect on fertilizer use efficiency. Where, the crop is taken under rainfed condition, the moisture stress at early growth stages or excess rains after planting have a direct bearing on best utilization of applied nutrient. This is true for K, a highly mobile nutrient prone to leaching to lower horizons in light soils or it may be washed out under surface run off. Several approaches tried in increasing the efficacy of P and K fertilizers in potato crop have been found to be quite effective.

Plant needs most of P at early stages of development thus its application at planting is considered more useful. Moreover, because of its low mobility and P fixation in acidic and laterite soils, its placement near the active root zone or near the seed is recommended as this gives *better* results than application of P above or below seed. Its furrow placement has been more economical than broadcasting or band placement in acidic hill soils of Shimla and Shillong. Like P, application of K at planting time in furrows is better than its application above or below seed or broadcasting in hill soils.

Effect of continuous P application: Continuous application of P to all the crops in sequence may lead to build up of in the soil. Only partial removal by crop uptake than applied P is attributed for build up of P in soils. High amount of P in soil tends to decrease the yield in alluvial soils. Excess P in soil decreases the availability of Zn, Mn, Fe and Cu in soil by forming insoluble phosphates with these elements and yield of many crops have been reported to decrease.

Effect of continuous K application: No deleterious effect of continuous application of K is reported. Reversible reaction between available, exchangeable and fixed K fraction in soils is attributed for maintenance of available K levels in the soil. Secondly, the uptake of K from soil is often in excess of the applied K resulting in negative balance of K in the soils.

Residual effects of P and K fertilizers: Major part of applied P is transformed to inorganic Al-P, Fe-P and Ca-P depending upon the soil reaction and is not immediately available to the plant. Therefore continuous application of P over a time leads to accumulation of P in the soil thereby raising available P status of the soil. Among the inorganic and organic sources of P and K, the residual effect of FYM has been found to be higher than that of fertilizer P or K. The soil analysis also indicate the superiority of organic manures like FYM over P and K fertilizers in improving the available P and K content in soils.

Diagnosis of micronutrient deficiencies plants

Accurate diagnosis is the first step for precise monitoring and efficient correction of micronutrient disorders in vegetable plants and soils. Several approaches have been in use to diagnose the magnitude of micronutrient deficiencies in vegetable plants and soils and are discussed hereafter (Fig. 7).

Visual diagnosis: Signs of stress will appear whenever any one of the micronutrients becomes severely deficient beyond the "marginal deficiency stage" of development. Each element develops its own characteristic stress symptoms. A visual diagnosis of symptoms should be made as soon as possible after the symptoms have appeared. Later on, the symptoms can not be easily classified because they will be modified by other environmental factors.

Deficiency symptoms

Zinc: Zinc deficiency of vegetable, often known as fern leaf or little leaf, appears on young developing leaves. Deficient plants show severe stunting and bronzing or yellowing of the foliage, usually around the leaf margins, starting from the tips. Youngest leaves are cupped upwards and rolled to such an extent that the terminal growth esembles that of ferns. Leaves of affected plants are smaller and their upper internodes are shorter.

Iron: Its deficiency appears initially as a yellowing of young leaves near the growing point of the plant. With time, the leaves become light yellow to nearly white. During the deficiency stage, blade tips remain green for longer time. Netted green veining occurs when traces of iron are absorbed and translocated along the veins for chlorophyll formation. Green veining is actually an iron recovery symptom.

Manganese: The first sign of its deficiency is yellowing and slight cupping of younger leaves. Pinkish colour develops at the base of younger chlorotic leaves and relatively old leaves show dark to black spots. With increased deficiency, dark to black spotting develops between the veins with a large increase in spotting, appearing along larger veins and the mid rib. These symptoms, darkening and cupping, increase in severity with time. During mild deficiency, upper parts of the plants become somewhat chlorotic but do not develop dead spots.

Copper: An early sign of its deficiency is the development of a uniform, light green colour of young, immature leaf blades similar to those of molybdenum, manganese and iron deficiencies. Thereafter, its deficiency is manifested primarily by a pronounced upward cupping and inward rolling of the young, relatively large, leaf blades. This is a sharp contrast to the small, narrow leaf blades of zinc deficiency.

Boron: Its deficiency causes the formation of a bushy plant with droopy leaves. Blades crinkle, cup upwards and are bordered by light brown tissue. Its deficiency, like calcium deficiency, affects the growing points. Immature centre leaves become deformed and the growing point dies. In case of mild boron deficiency, slight upward curling of the margins of the older leaves is visible.

***Molybdenum*:** The symptoms of its deficiency are marked chlorosis, associated with reduction in growth and yield.

Fig. 7 : Deficiency systems pattern in plant

Nutrient status in Indian soils

Essential nutrients are irreplaceable by other elements, which are directly involved in plant metabolism and without it, plant cannot compete their life cycle. For field crops these include Fe, Mn, Zn, Cu, B, Mo and Cl. Although, Ni, Si and Co not essential as per essentiality criteria (Arnon and Stout, 1939), but are beneficial to the plants. The essential micronutrients are metals (except B and Cl) and its uptake is affected by soil (Lindsay, 1991; Stevenson, 1986; Lake *et al*., 1984), plant (Barber, 1995; Marschner, 1995) microbial and environmental factors (Romheld and Marschner, 1986; Clark and Zeto, 2000). Research on micronutrient requirement of crop plants in India was initiated during late 1950's and intensified after the inception of AICRP on 'Micronutrients of Soils' during 1966-67. The reviews on the micronutrient research in India by Kanwar and

Randhawa (1967), Agarwala and Sharma (1979), Takkar *et al.* (1989), Sharma *et al.* (1996), were highlighted on the identification and correction of micronutrient disorders.

The fertility problem and nutrient deficiencies in Indian soil has progressed over the time (Chhonkar and Rattan, 2000). Emerging Multi-Nutrient Deficiencies in Soils is given Fig. 8 & 9. Indian soils are generally deficient in nitrogen, phosphorus and marginally adequate in potassium. Sulphur has become critical on low organic matter coarse-textured soils. Zinc deficiencies are widespread across the country (Fig. 10). Magnitude of boron deficiencies is next only to zinc (Fig. 11). Manganese deficiencies are common on coarse textured. The soil fertility map depicting the micronutrient status of Indian soils are cited (Kanwar, 1976; De *et al.,* 2004).

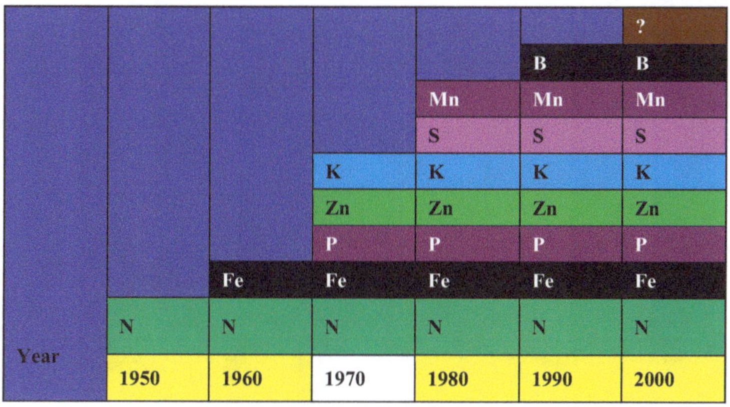

Fig. 8 : Emerging Multi-Nutrient Deficiencies in Soils

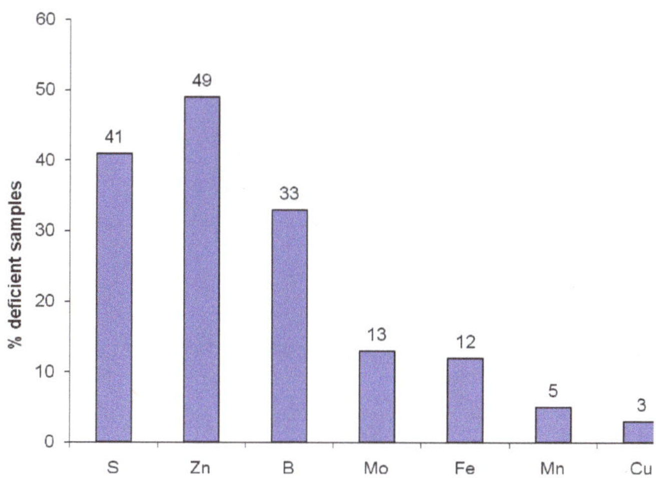

Fig. 9 : Nutrient Status- Micro and Secondary deficiency in India

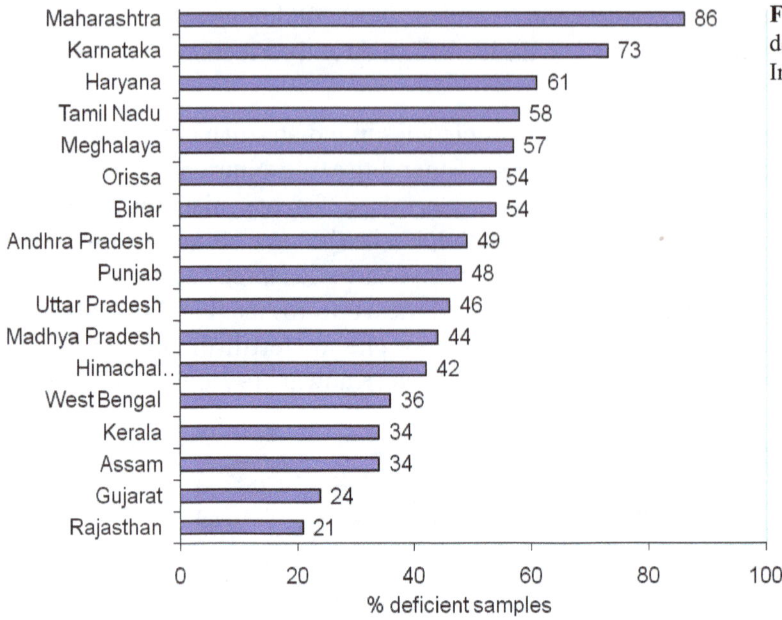

Fig. 10 : Zinc deficiency in India

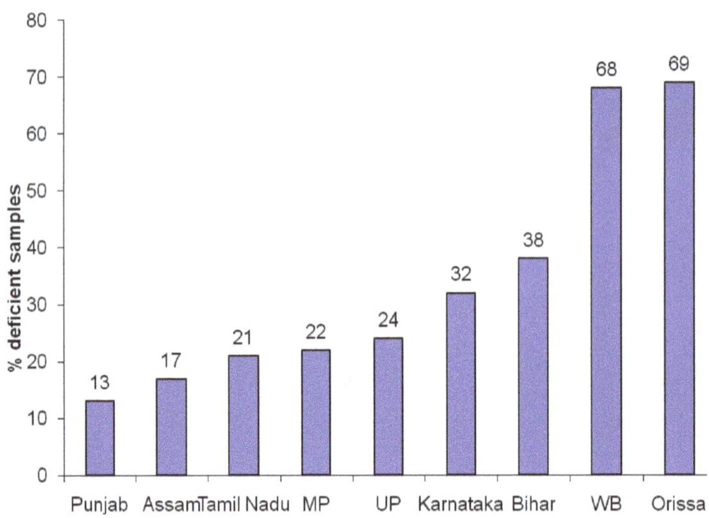

Fig. 11 : Boron deficiency in India

Micronutrients, which include zinc, iron, manganese, copper, molybdenum, boron and chlorine, play a specific role in the growth and development of a plant. Even though these elements are needed in only minute traces, many soils do not supply them in sufficient quantity for healthy growth and optimum yield of vegetable crops. The use of pure NPK fertilizers and high yielding vegetable crops varieties with increased nutrient demands, the decreasing availability of farmyard manure in areas where mixed farming has given way to intensive arable agriculture, have combined to increase the demands made on the soil in terms of its ability to supply micronutrients to plants. It is becoming evident that without the use of some of the micronutrients, it is not possible to get the maximum benefit of other inputs.

Soil properties of the Thar Desert

The soil of the Thar Desert belongs to the order Aridisols and is sandy desertic, very poor in organic carbon, water holding capacity and deficient in zinc. The knowledge of soil properties helps the vegetable growers to raise a successful crop. The physico-chemical properties of desert soil are as follows (table 3a&b):

Table 3a : Physical properties

Texture	Loamy sand
Sand content	80-90 %
Silt content	2-5 %
Clay content	2-10 %
Structure	Single grain structure
Bulk density	1.58-1.60 mega gram/cubic meter
Soil moisture retention capacity	3.0-4.5% (w/w)
Wilting point	1.5-2.5%

Table 3b : Chemical properties

pH	7.5-8.5	Slightly alkaline
Ec	0.16-0.21 dsm^{-1}	Non-saline
Organic carbon	0.03-0.08 %	Low
Available N	65-100 kg/ha	Low
Available P_2O_5	5.0-8.0 kg/ha	Low
Available K_2O	200-480 kg/ha	High
Available Fe	5.0-6.0 mg/kg	Sufficient
Available Mn	4.0-4.5 mg/kg	-do-
Available Cu	0.10-0.1 8 mg/kg	-do-
Available Zn	0.10-0.15 mg/kg	Deficient

In Rajasthan, 88, 45 and 34% soils are deficient in nitrogen, phosphorus and sulpher (Fig. 12). Zinc deficiency is widespread throughout India including soils of the Thar Desert. The magnitude of boron deficiency is next to zinc. The soil health and organic content is not good and also deteriorating fast. The continuous deterioration of soil health does not allow harvesting optimum yield of vegetable crops. Management of micronutrients is a key factor in determining the production and productivity of vegetables in the era of intensive vegetable culture. Further the deficiency of micronutrients, if not properly taken care would threat the ecological sustainability of intensive cultivation of vegetable crops. A detailed soil map with reference to micronutrient status of arid region soils is available in literature. However the information regarding micronutrient status of desert soils are scare and the soil map of Bikaner and Jaisalmer districts is as follows.

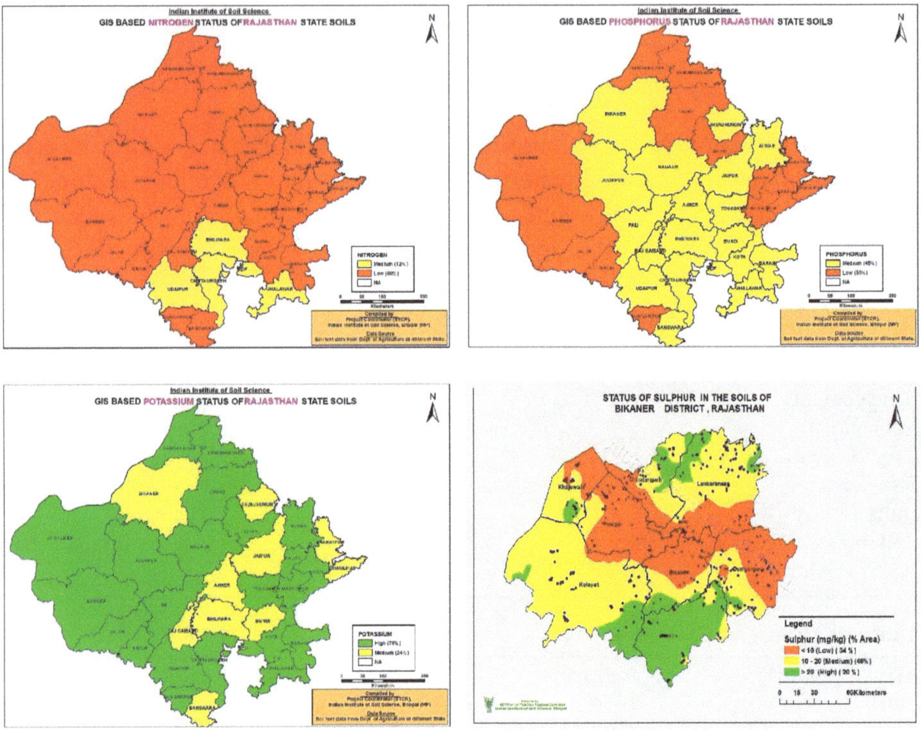

Fig. 12 : NPK and & S status in Rajasthan

The crop production in the hot arid regions are constrained by low and erratic rainfall, high evapo-transpiration and adverse soil physical and fertility conditions. Arid region soils are low in organic matter, macronutrient and micronutrient (Fig. 13&14) (Shyampura *et al.,* 2002, Rathore, 2009, Yadav and Meena, 2009, Yadav, 2011, Chattopadhyay *et al.*, 1997, Singh, 2006, Singh, 2008). The low organic matter has been attributed to high temperature, low rainfall, scanty vegetation and single grained texture of soil.

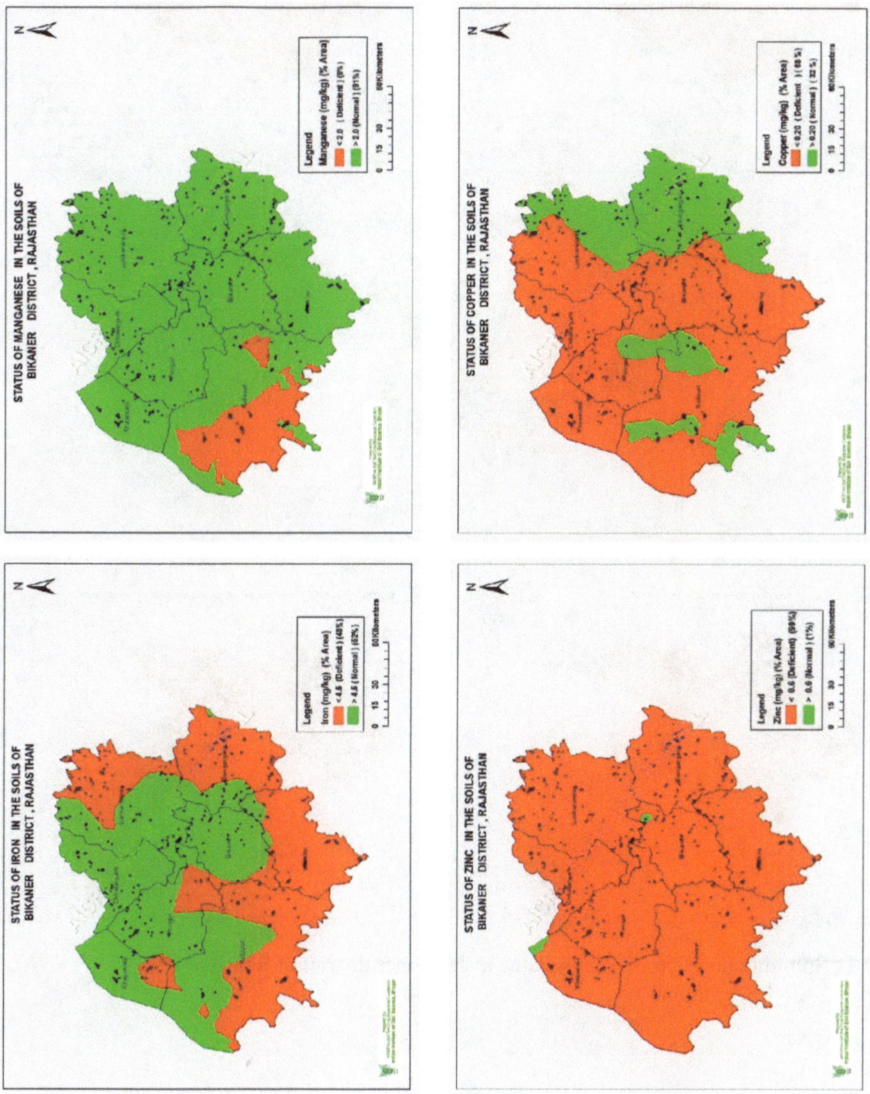

Fig. 13 : Sulpher and micronutrients status in Bikaner district of Rajasthan.

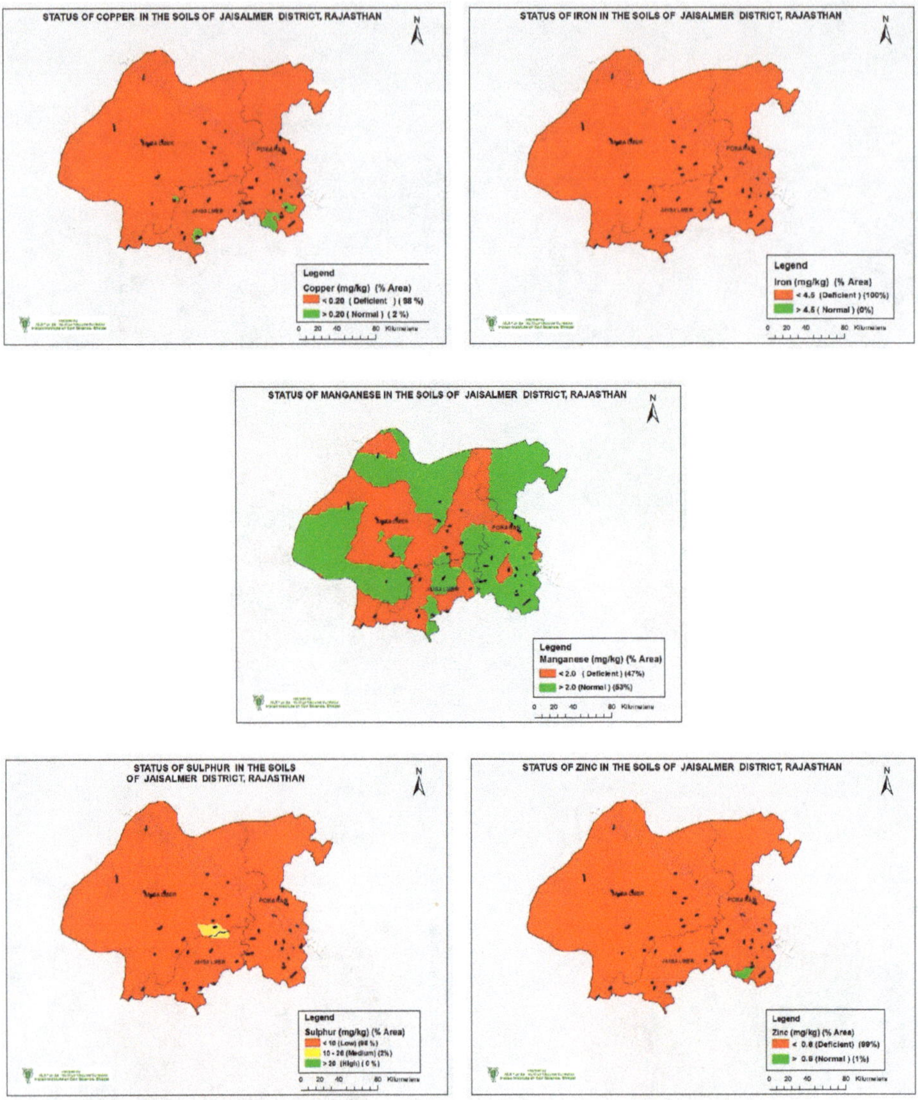

Fig. 14 : Sulphur and micronutrients status in Jaisalmer district of Rajasthan.
Source: ICAR-IISS, Bhopal

Removal of NPK and micronutrients from soil

Many horticultural crops are heavy removers of nutrients and high yields can only be sustained through the application of optimal doses in a balanced proportion. Approximate quantities of nutrients removed by some important horticultural crops are given in Table 4. Depleted plant nutrients are essentially supplied through manures and fertilizers. Nutrient management for vegetable crops is complex, requiring the integration of biological, chemical, and economic factors. The global food requirement continues to grow, and fertilizer is presently estimated to be responsible for at least half of the current food supply (Eriman *et al.,* 2008, Stewart *et al.,* 2005). Responsible nutrient management production must include consideration of environmental, economical, and social components. There is a need of integrated, balanced and effective fertilizer management to take care of proper replenishment and compensation of nutrient losses from soil and locked-up nutrients in the growth of horticultural crops.

Table 4 : Nutrient removal by different horticultural crops

Crop	Yield (t/ha)	Nutrient removal (Kg/ha.)		
		N	P_2O_5	K_2O
Vegetables				
Potato	28.0	202.0	50.0	225.0
Brinjal	60.0	175.0	40.0	300.0
Tomato	37.0	104.0	22.0	141.0
Cauliflower	50.0	250.0	100.0	350.0
Cabbage	37.0	112.0	28.0	112.0
Beans	25.0	130.0	40.0	160.0
Green peas	25.0	55.0	20.0	40.0
Lettuce	30.0	107.0	30.0	234.0
Spinach	25.0	120.0	45.0	200.0
Celery	20.0	140.0	55.0	220.0
Onion	30.0	73.0	36.0	68.0

Source: Ganeshamurthy *et al.,* 2011

Nutrient Management

Nutrient management is the science and art directed to link soil, crop, weather, and hydrologic factors with cultural, irrigation, and soil and water conservation practices to achieve the goals of optimizing nutrient use efficiency, yields, crop quality, and economic returns, while reducing off-site transport of nutrients that may impact the environment (Delgado and Lemunyon, 2006). Nutrient management, particularly in horticultural crops, is the skillful task of matching a specific field, soil, climate, and crop management conditions to rate, source, timing, and place (commonly known as the nutrient stewardship) of nutrient app lication (Robert L Mikkelsen, 2011). Some important factors that need to be onsidered when managing nutrients include (a) the application of nutrients considering crop quality and achievable optimum yields; (b) application, and timing and management of nutrients using a budget based on all sources and sinks active at the site; and (c) the management of soil, water, and crop to minimize the off-site transport of nutrients from nutrient leaching out of the root zone, surface runoff and volatilization (or other gas exchanges). There can be potential interactions because of differences in nutrient pathways and dynamics. For instance, practices that reduce the off-site surface transport of a given nutrient may increase the leaching losses of other nutrients. These complex dynamics present nutrient managers with the difficult task of integrating soil, crop, weather, hydrology, and management practices to achieve the best balance for maximizing profits while contributing to the conservation of our biosphere. Quality consciousness of the humans for quality produces, particularly of horticultural crops, is increasing. As a result one important aspect of crop production is that of crop nutrition and quality and high productivity can only be achieved if plants are properly fed. Recommendations for fertilizing fruits, vegetables and other horticultural crops are based in part on soil test results. Soil testing provides information on lime and Ferilizer needs prior to planting and is particularly well calibrated for nutrients such as phosphorus, potassium, magnesium, calcium, sulphur, zinc, and boron. Soil testing prior to planting takes the guesswork out of making fertilizer recommendations and leads to more efficient nutrient management. However, for fine tuning ferilizer needs of the perennial horticultural crops, a combination of soil testing and tissue analysis should be used (Ganeshmurthy *et al.*, 2015 and Sharma *et al.*, 2019).

Tissue testing in vegetable crops

Soil tests, although useful in predicting fertilizer and soil amendment needs, are not the final measure of what nutrients a plant will absorb. Because temperature, moisture regimes, soil acidity and other soil conditions may modify the uptake of different nutrients by plants, it is sometimes necessary to determine nutrient content in the plants to evaluate the actual soil nutrient availability status. Total plant analysis and green tissue analysis are the most important diagnostic techniques for determining deficient, sufficient or excessive amounts of essential elements in plant tissue. Tissue tests are of immense help to farmers in overcoming many problems and are the better approaches for nutrient management of horticultural crops. The success of this approach depends mainly on selecting the right tissue and deciding on the right stage of sampling (Bhargawa and Chadha, 1993). The sampling guide and the optimum nutrient levels developed for different vegetable crops are given in Tables 5 and 6 respectively. Direct assessment of nutritional status of horticultural crops through chemical analysis of the plants may be the effective means of persuading farmers to rationalize ferilizer applications. Unfortunately in the past the results of leaf analysis have proved difficult to interpret because of the changes in the nutrient concentration with the age of the plant and also changes in the concentration of other nutrients (Kenworthy, 1964 and Reuter and Robinson, 1986). However, new methods of interpretation of leaf analysis have overcome these difficulties. Based on well-designed long-term experiments and survey of cropping enterprises the leaf nutrient norms for selected crops were developed. Common nutritional disorders in horticultural crops are given in Table 7.

Table 5 : Plant tissue sampling guidelines for different vegetables

Crop	Plant part/Index tissue	Growth stage/time
Bean	Upper most recent fully developed trifoliate leaves	-
Brinjal	Leaf blades with midribs minus petioles from most recently fully developed leaves	Active growth
Cabbage	Wrapper leaf	60 Days
Carrot	Most recent fully matured leaf	Mid-grown
Cauliflower	Most recent fully matured leaf	At heading
Cucumber	5th leaf from tip	Flowering to fruit set
Cluster bean	1st fully developed leaf	
Garlic	Most recent fully developed leaf, no white portions	Pre-bulb
Muskmelon	5th leaf from tip	Flowering to fruit set
Onion	Top no white portions	1/3 to 1/2 grown
Pea	Leaflets from most recent fully developed leaflets	First bloom
Potato	Most recent fully developed leaf	Half grown
Radish	Most recent fully developed leaf	30 days
Spinach	Most recent fully developed leaf	30-50 days
Tomato	Leaf adjacent to infloresc	Mid-bloom
Watermelon	-do-	Flowering to fruit set

Source: Bhargava and Chadha, 1993; Raghupati *et al.,* 2014

Table 6 : Index tissue sampling technique and optimum range of nutrient requirement in vegetable crops

Crop	Sampling time/part	Optimum leaf nutrient norms		
		N (%)	P (%)	K (%)
Tomato	Petioles opposite or below top flower cluster at pre-bloom	3.05-5.00	0.71-1.30	4.00-8.00
Turnip	Most recently fully developed leaf	3.50-5.00	0.33-0.60	3.50-5.00
Cabbage	Top plants 2-6 week old	3.00-5.00	0.33-0.75	3.50-5.50
Carrot	Most recently fully matured leaf at mid growth	3.00-5.00	0.20-0.40	2.80-4.00
Onion	Tap, No white portion one/third to half growth	5.00-6.00	0.30-1.20	3.50-5.00
Brinjal	Leaf blade with midribs minus petiole most recently developed leaves	3.00-5.00	0.33-0.75	3.00-5.50
Cucumber	Leaf blade with midrib, fifth leaf from tip omit unfurled leaves, flower bud start to small fruit	4.30-6.00	0.25-1.00	3.50-4.50
Pepper	Most recent fully developed leaf, Flower bud start to small fruit	4.00-6.00	0.35-1.00	4.00-6.00
Redish	Most recent fully developed leaf	3.00-6.00	0.30-0.70	4.00-6.00
Watermelon	Fifth leaf from tip, omit unfurled, flower start, to small fruit	4.00-5.50	0.30-0.80	4.00-5.00
Pea	Recently fully developed leaflets, first bloom	4.00-6.00	0.30-0.80	2.00-3.50

Table 7 : Common nutritional disorders in horticultural crops

Deficient element	Disorder	Crop	Symptoms
Boron (B)	Brown heart	Turnip	Rotting of centre of root
	Heart rot	Beets	Death of centre of crown, rotting of centre of root
	Hollow stem	Cauliflower	Rotting of centre of stem
	Bitter pit	Apple	Decay or corking of the flesh under the skin
Calcium (Ca)	Blossom-end rot	Tomato, Pepper	Decay of the blossom-end of the fruit
Manganese (Mn)	Speckled yellows	Sugar beet	Chlorosis between leaf veins, inward curling of leaves
	Marsh spot	Pea	Brown area in centre of seed
	Little leaf	Apple	Small, malformed leaves shortened internodes
Molybdenum (Mo)	Whiptail	Cauliflower	Reduction or suppression of leaf

Micronutrients in soil-plant system

The amount and distribution of micronutrients in soils are influenced by parent materials, levels and form of soil organic matter, pH, Eh, mineralogy, particle size distribution, soil horizon, soil age, soil formation processes, drainage, vegetation, anthropogenic and natural processes and microbial population (Baligar *et al.,* 1998; Stevenson, 1986; Tate, 1987 and Fageria *et al.,* 2002). Comprehensive information for optimum time of sample collection, selection of index tissue and growth stage for plant analysis is extremely important (Mortvedt, 1991; Bergman, 1976). If samples are collected earlier than the optimum crop growth stages stated in the Table 3, the levels of some elements, e.g. nitrogen, phosphorus, partly potassium, manganese may be higher, and those of other e.g. calcium in leaves, magnesium and boron will be lower than given. This situation will be reversed if samples are taken later. Apart from this, varietal differences can cause variations of "adequate ranges" within certain limits. Plant tissue analysis aids in the characterization of soil chemical properties and mineral nutrition of plants. A wet oxidation procedure results in the conversion of Na, K, Ca, Mg, Mn, Fe, P, Cu, Zn and others to proper form for analytical determination (Page *et al.,* 1982; Jackson, 1967). Dry assay is also useful for mineral analysis in plant ash. An understanding of the nutrient content associated with latent ("hidden hunger") or acute deficiency, luxury consumption and possible toxicity disorder is essential when determining nutrient deficiencies or excesses by plant or leaf analysis. The "adequate or sufficient range" is of greatest interest as there will be neither a deficiency nor an excess of plant nutrients (Table 8). Adequate ranges of micronutrients in dry matter (DM) of various vegetable crops (Table 9) will be an important aid to diagnose the deficiency and toxicity (De and Rai, 2005).

Table 8 : Ranges of micronutrients concentration in plants and soils

Elements	Concentration range in plants (ppm) (1)			Concentration in soil (ppm) (2), (3)	
	Critical	Sufficient	Toxic	Normal	Critical
Fe	<50	50-250	>1000	200-500,000 or (3.8%)	-
Mn	10-20	20-300	300-500	7-10,000	1000-3000
Zn	15-20	20-100	100-400	1-900	70-400
Cu	3-5	5-20	20-100	2-250	60-125
B	<10	10-100	50-200	2-150	15-25
Mo	<0.1	0.1-0.5	10-50	0.1-40	2-13
Cl	<2000	2000-20,000	>20,000	20-900	70-200

(1) Bennett (1993) and McBride (1994); (2) Alloway (1995a), Cox (1987), Martens and Lindsay (1990), Sims (2000), Sims and Johnson (1991); (3) Kabata-Pendias and Pendias (1992).

Table 9 : Mineral nutrient contents in dry matter of vegetable crops.

Crop	Adequate concentration range (ppm)					
	Fe	Mn	Zn	Cu	B	Mo
Beet	50-200	35-100	20-80	6-15	35-100	0.2-1.0
Brinjal	40-200	30-125	20-60	5-15	25-75	0.2-0.8
Cabbage	30-200	30-100	20-60	5-12	25-80	0.4-0.7
Carrot	50-300	50-100	30-80	7-15	30-80	0.5-1.5
Cauliflower	30-200	30-100	30-70	5-12	30-80	05.-1.0
Chilli	30-300	30-100	20-60	8-15	40-80	0.2-0.6
Cowpea	50-250	400-425	40-50	5-7	150-200	0.2-0.3
Cucumber	50-300	60-120	35-80	7-15	40-80	0.8-2.0
Muskmelon	50-100	40-10	20-60	5-12	25-60	0.2-1.0
Onion	60-300	40-100	20-70	7-15	30-50	0.15-0.3
Peas	50-300	30-100	25-70	7-15	30-70	0.4-1.0
Potato	40-100	40-100	20-80	7-15	25-70	0.2-0.5
Tomato	40-200	40-100	20-70	7-15	40-80	0.3-1.0
Water melon	50-300	30-100	20-70	5-10	30-80	0.2-1.0

*Mortvedt *et al.*,1991; Mengel and Kirkby,1982;Jackson, 1967; Jones *et al.*, 1990; Bergmen,1976

Micronutrient management

The plant nutrient content has been correlated with soil content, yield (Lundegardh, 1951; Tandon, 1991) and quality factors such as carotene (Jackson, 1967). Micronutrient disorders can occur when the nutrients are not present in the soil system or present but unavailable to the plant. Occasionally, plants with roots damaged by Pythium or other pathogens may show micronutrient deficiency symptoms. Toxicity can occur when micronutrients are applied in excess. Toxicity symptoms are difficult to recognize visually and are usually mistaken by deficiency symptoms. Sometimes stresses, such as drought, heat, over watering, or mechanical injury cause development of symptoms that are mistaken for nutrient deficiencies.

Functions of micronutrients

Micronutrients serve mainly as constituents of prosthetic groups in metalloproteins and as activators of enzyme reactions. The function of micronutrients has been reviewed extensively by Marschner, 1986; Romheld and Marschner, 1991; Mortvedt *et al.,* 1991; Mortvedt 2000 and Fageria *et al.,* 2002.

Factors affecting micronutrient availability

Soil factors like pH, organic matter content, cation exchange capacity (CEC), clay content, temperature and moisture affect the availability of micronutrients to crop plants. Soil pH influences, concentration of ions in solution and solubility, mobility and acquisition of elements by plants (Fageria *et al.,* 1997). The availability of B, Cu, Fe, Mn and Zn usually decreases and Mo increases as soil pH increases. Soil organic matters (OM) are grouped into water insoluble (humic acid/ humin) and water soluble (fulvic acid) compounds. Soil OM content has been related to increased, decreased and no effects on micronutrients availability to crop plants (Stevenson, 1982; Tisdale *et al.,* 1985; Mortvedt, 2000). Micronutrients availability decreases at low temperatures and moisture content because of reduced root activity, low rates of dissolution and diffusion of nutrients. Nutrient supplies to plants are associated with factors like concentration of nutrients inside plants and in soil solution, supply and chemistry of root surface/ rhizosphere, chemical equilibria and interactions of one nutrient with another. Information on mechanism, process and factors associated with mineral nutrient uptake and translocation has been reported earlier (Barber, 1995; Marschner, 1995 and Lindsay, 1991). Plant factors such as root morphology (length, density, volume, surface area), root induced changes (secretion of H^+, OH^-, HCO^{-3} ions), root exudation of organic acids (citric, malic, tartaric, oxalic, phenolic), sugars, amino acids (phytosiderophores), secretion of enzymes (phosphatase), plant demand, plant species/ cultivars and microbial associations (release of CO_2, PGR, enzymes, rhizobia, mycorrhizae etc.) have profound influences on plant ability to absorb and utilize micronutrients from soil. Micronutrient uptake by roots depends on nutrient concentrations around root surfaces, root absorption capacity and plant demand. Nutrient uptake by plant is a dynamic process in which nutrients are continuously replenished in soil solution from the soil solid phase, and then transported to roots as uptake proceeds. In soil system nutrients move to plant roots by (i) mass flow (ii) diffusion and (iii) root interception (Barber, 1995). More than 90-95% of N, B, Cu, Zn and considerable quantity of Fe (65%) supplied to plants by mass flow.

Considerable quantities (>80% incase of P, K) of Mn, B and Fe (>20%) move by diffusion. Root interception can provide significant amount of plant requirements for B, Zn and Mn. Micronutrient uptake by vegetable crops involves: (i) Selectivity-where certain nutrients are absorbed preferentially over others. (ii) Accumulation-where certain minerals accumulate at higher concentrations in cell sap than in external soil solution. (iii) Genotypes-where distinct differences exists among plant species and within species (Epstein,1972; Mengel and Kirkby, 1982; Romheld, 1991; Marschner, 1995; Sumner, 2000). The micronutrient interactions among the various nutrients also influences and changes in nutrient concentrations in plants as well as growth responses (Olsen, 1972; Morghan and Mascagm, 1991, Fageria, 2002). The concentration of one micronutrient in soil plant system may affect the level of other micronutrients through a process called antagonism. For example, too much iron may produce manganese and zinc deficiencies, while high levels of manganese may result in iron and zinc deficiencies. Copper and zinc are also antagonistic (Table 10).

Table 10 : Availability of micronutrients as affected by antagonism in soil.

Element	Availability reduced by
Boron	Organic nitrogenous fertilizer and high levels of phosphorus
Copper	High levels of zinc, nitrogen and phosphorus.
Iron	High levels of copper, manganese and zinc and phosphorus
Manganese	High levels of potassium, phosphorus, iron, copper and zinc.
Molybdenum	High levels of manganese and nitrate nitrogen fertilizers
Zinc	High levels of copper and phosphorus

Characteristic deficiency symptoms of micronutrients

Reviews on deficiency symptoms of micronutrients in crop plants and its management have been focused mostly in cereals, pulses and oilseed crops (Agarwala and Sharma, 1979; Takkar *et al.,* 1989; Sharma *et al.,* 1996). Recently micronutrient disorders in vegetable crops and their management have been discussed at length (De *et al.,* 2004, 2004a).

Correcting deficiencies

A wide variability in concentrations of nutrients for soil and foliar spray are reported in the literature. Several factors associated with foliar applications include plant (leaf age, species, nature, i.e. thick or thin leaves, nutritional status and requirements), climatic (light, temperature, humidity), chemical (form, carrier, adjuvant) influence the effectiveness of foliar spray (Kannan, 1990; Mortvedt, 2000). Further nutrient penetration rate in the leaf surface, run off from hydrophobic surfaces or rate of washed off nutrients by rain or spray water, rate of drying of spray solution, rate of translocation from leaf surface to other plant parts determines the spray effectiveness (Horesh and Leavy, 1981). This variation besides the factor listed above further influenced by soil type, crop species/ cultivars, desired crop yield and climatic conditions (Martens and Westermann, 1991; Mortvedt, 1991, 1994, 2000). The concentrations of micronutrient for correcting deficiencies in vegetable crops are summarized (Table 11).

Table 11 : Application of micronutrient in vegetable crops

Element	Corrective measures	
	Soil application (kg/ha)	**Foliar application**
Fe	Annual treatment of 0.5-10 kg/ha $FeSO_4$	One spray of 2% $FeSO_4.7H_2O$, or 0.02-0.05% Fe-EDTA (Several sprays)
Mn	Generally not recommended	0.1 % $MnSO_4.H_2O$
Zn	0.5-35 Kg. $ZnSO_4$	0.1 -0.5 % $ZnSO_4.$ $7H_2O$ Solution
Cu	1-20 Kg. $CuSO_4$ ha^{-1}(Every 5-10 years)	0.1 – 0.2 % solution of $CuSO_4.5H_2O$
B	0.25-7 kg borax ha^{-1} Soil application Preferred	0.1-0.25 % B solution
Mo	0.01-1 kg Na or NH_4 Molybdate ha^{-1}	0.07-0.1% Na or NH_4 molybdate
Cl	20-50 kg KCl ha^{-1}	Unknown

Source: Bould *et al.,* 1983; Fageria *et al.,* 1977, 2002 and Watermann, 1991.

Deficiencies can be corrected by adding the micronutrient that is in deficit or by correcting the factor for its unavailability. If only one micronutrient is deficient, one should not apply a micronutrient complex fertilizer because, as mentioned above, imbalances can cause antagonism. Micronutrients can be applied (Tables 11&12) in three different ways: i) soil application ii) applied as a single foliar spray at just before flowering stage, suspected mild deficiency and iii) applied as a foliar spray at critical crop growth stages (i.e. active growth, flowering, fruit setting), suspected severe deficiency. Boron is usually applied at 0.25 to 3 kg/ ha, higher rates for broadcasting and lower rates for band placement. Foliar spray of boric acid/borax varying concentration from 0.5-1 ppm (Sindhu *et al.,* 1989, 1993; Bhonde *et al.,* 1995) to 0.1-0.25% i.e. 1000-2500 ppm (Fageria *et al.,* 2002) has been evident in literature. Foliar application of 3000 or 5000 ppm of Zn, Cu, Fe and Mn in tomato extended flowering period, reduced the number of flower cluster by up to 20%, fruit set up to 35% and yield/ha, while soil application @ 2.5-10 kg/ha was found beneficial (Mallick and Muthukrishnan, 1980). Yield in okra cultivars cv. Pusa Sawani were significantly increased with foliar application of aqueous solutions of mixed chloride, cobalt chloride and ammonium molybdate @ 400, 200 and 200 ppm at flowering/fruiting stages (Tosh *et al.*, 1980). Foliar sprays of 0.125-4% boric acid at 15-day intervals after 7 days of transplanting to curd formation in cauliflower or soil application @1.5 kg /ha boric acid was optimum for high curd yield (Kotur, 1998). The soil application rates of $CuSO_4$ required to correct deficiency have been reported as low as 1.1 kg ha^{-1} to 30 kg ha^{-1} for vegetables (Rao and Deshpande, 1971; Singh and Tiwari, 1995; Reddy *et al.,* 1985). Soil application of $CuSO_4$ is usually more effective than foliar sprays (Karamanos *et al.,* 1986, Solberg *et al.,* 1993). The concentrations of Cu for foliar spray are varying from 1 ppm in onion (Bhonde, *et al.,* 1995, Sindhu *et al.,* 1989, 1993) to as high as 5000 ppm in tomato (Mallick and Muthukrishnan, 1980; 1980a). Mainly foliar sprays correct Fe deficiency where soil applications are generally ineffective. The concentrations of Fe vary from 50 to 5000 ppm in different vegetable crops (Table 8). The soil application rate for $FeSO_4$ varies from 5-30 kg ha^{-1}. Both soil application and foliar spray of $MnSO_4$ were effective for correcting deficiency in vegetable crops. Concentrations ranging from 10 ppm to 5000 ppm for foliar spray and soil application from 5 to 25 kg ha^{-1} is common for Mn. Zinc deficiency can be corrected by either foliar or soil application of $ZnSO_4$ @ 5 to 5000 ppm or 5-25 kg ha^{-1} as indicated in literature (Table 12). Foliar applications of Mo have usually been more effective than soil applications for vegetable crops. The applications of 0.01 to 0.5 kg Mo ha^{-1} will generally correct Mo deficiency. The requirement of Mo is quite low in vegetable crops when compared to other micronutrient elements.

Improving micronutrients use efficiency

In recent years, the increased anthropogenic activity (burning fossil fuels, application of sewage and industrial sludge), use of amendments (fertilizers, manures, lime), application of pesticides and deposition of atmospheric particles (Alloway, 1995) induces addition of toxic trace elements like Cd, Cr, Ni, Pb, Cu, Zn, As, Co and Mn (some are essential micronutrients) in soil-plant system. Excessive levels of trace elements may pose phytotoxicities to vegetable plants. Organic matter helps in maintaining good soil structure, increases water holding capacity, exchangeable ions, reduces leaching of nutrients and toxicities. The addition of crop residue, green manures, composts, animal manures, growing cover crops, reducing tillage and avoiding burning of crop residues can significantly lead to improved plant growth and acquisition of micronutrients (Fageria *et al.*, 2002). The accumulation of micronutrients varies among plant species and cultivars/genotypes within species (Marschner, 1995), which have been attributed by genetics, environment, physiological/ biochemical mechanism, response to agronomic management practices and tolerance to pest and diseases. Genetic variations in plant acquisition of micronutrients have been reviewed (Brown *et al.*, 1972; Duncan, 1994; Duncan and Carrow, 1999; Gerloff and Gabelman, 1983; Graham, 1984; Marschner, 1995). The development of vegetable genotypes effective in the acquisition and use of micronutrients.

Role of micronutrients in plant health

Micronutrient concentrations in plants are important in host ability to resist or tolerate infectious pathogens. Plant nutrition is an important component of disease control (Huber and Wilhetm, 1988). The deficiencies of Cu, Fe, Mn, Mo and Zn reduced growth and sporulation of the Fusarium oxysporum (Jones *et al.*, 1990). Cu has been used extensively over time as a fungicide and suppresses many soil borne diseases. Boron sufficiency in plants reduces the incidence and severity

Table 12 : Application rate of micro-nutrient for correcting deficiencies in vegetable crops.

Crop	Foliar spray (ppm)						Soil application (Kg/ha.)	References
	Fe	Mn	Zn	Cu	B	Mo		
Onion	50-150	0.25	3-5	1-3	0.25-1.0	1-4	Cu 13.4 B 1.8	*Rao & Deshpande*,1971 *Singh and Tiwari*,1995, *Sindhu et al*,1989,1993; *Bhonde et al.*1995; *Sliman et al*, 1999
Tomato	80-200	100-300	30-100	7-20	30-70	<0.4	Cu 10-30, B 5-15	*Arora et al.*, 1983, *Reddy et al.*, 1985, *De et al.* 2004
	3000-5000	3000-5000	3000-5000	3000-5000	200	25	Zn 10-20, Mn 10-20 Cu 5-15, B 5-15	*Mallick and Muthu- krishnan*, 1980, 1980a, *Shafshak et al.*, 1984, *De et al.*, 2004
Okra	50-2000	100-3000	30-1000	5-200	30-70	NH4 molybdate 200-300; 1 to 4	Zn 5-25, Mn 10-30 Cu 5-15 B 5-15	*Tosh et al*, 1980, *Aduyi and Idowu* 1981, *De et al.*, 2004
Cucumber	100-200	100-300	40-100	8-20	40-80	Less than 0.5	-	*Aduyi and Idowu* 1981, *De et al.*, 2004
Beans	-	25-50	50-100	20-40	-	-	-	*Gabal et al.*,1985, *Jana et al.*, 1987.

of diseases, while B deficiency enhances them (Gupta, 1993; Graham and Webb, 1991). Manganese concentrations in host tissue commonly decrease as the incidence of disease increases. The severity of parasites on plants is influenced by many micronutrients (Engelhard, 1990; Graham and Webb, 1991; Huber, 1980). The lack of Zn, B, Mn, Mo, Ni, Cu and Fe in plant tissue can enhance various diseases on plants (Engelhard, 1990; Graham and Webb, 1991; Huber, 1980; Fageria, 1997; Baligar *et al.,* 1998). Mineral nutrients can affect factors by which plants resist pests. These factors are physical (surface properties, hairs and colour), mechanical (fibers, silicon) and chemical/biochemical stimulants, toxins, repellants (Marschner, 1995). Although, Si has not been discussed as a micronutrient, its importance in insect and disease resistance has been studied extensively in rice, sugarcane and cucumber (Menzies and Belanger, 1996; Savant, *et al.,* 1997; 1999).

Micronutrient and human health

More than 3 billion people of the world are anemic, 90% of them are in developing countries. Iodine deficiency is the leading cause of mental retardation in the world. Zinc deficiencies in pregnant women have been associated with child births. Supplementation and fortification of mineral nutrients appears to be a partial solution to the malnutrition problem. Biofortification through conventional breeding has so far proved to be of little or no significance (Misra *et al.,* 2004). A number of CGIAR organizations and national institutes are exploring the conventional breeding assisted by molecular markers tools to develop genotypes containing significantly higher amount of micronutrients. Rice, wheat and maize are among the world's most consumed staple foods. However, it would require a person to eat more than 5 kg of any of these staples each day to meet the requirements for vitamin A and iron (AVRDC, 2004). Consuming few spoonfuls of vegetable is a much more reasonable option for satisfying one's daily micronutrient requirements. Improving the genetic potential of vegetable crops to produce micronutrient rich diet can play an important role in tackling the malnutrition problem to a great extent.

Soil related problems under nutrient management strategies

Soil pH: Soil pH is an important chemical property that affects nutrient availability and microbial activity. In general, the optimum pH for most fruit and vegetable crops is between 5.8 and 7.0 for mineral soils and 5.4 and 6.2 for organic soils (peats and mucks). Two exceptions are lueberries and potatoes. Blueberries are adapted to acid soil conditions and grow best at a soil pH between 4.5 and 5.2. Potatoes can tolerate a wide range in soil pH; however, potato scab can become more of a problem in scab-susceptible varieties as soil pH increases above 5.3. As one of the most important soil chemical properties for optimal crop production, soil pH determines nutrient sufficiency, deficiency, toxicity, and need for liming (Fageria and Zimmermann, 1998) or addition of sulphur.

A pH range of 5.5 to 7.0 is suitable for most of the vegetable crops (Table 13). This pH range can assure high bioavailability of most of the nutrients essential for vegetable growth and development (Ronen, 2007). For example, at soil pH 8.0 or higher, iron and/or manganese bioavailability can't satisfy most vegetable crops' requirements. However, when soil pH reaches 5.0 or lower, solubility of aluminium, iron, manganese, and/ or zinc in soil increases and these elements become toxic to most vegetable crops (Osakia *et al.,* 1997).The list of vegetable crops along with their acidity tolerance are given in Table 14.

Table 13 : Soil pH range for optimal growth of selected vegetable crops

Vegetable Name	pH range	Vegetable crop	pH Range
Asparagus	6.0-8.0	Okra	6.0-8.0
Beans	6.0-7.0	Peas	6.0-7.0
Beet Root	6.0-7.5	Potato	4.5-6.5
Broccoli	6.0-7.0	Pumpkins	5.5-7.5
Cabbage	5.8-7.0	Radish	5.5-6.5
Carrot	5.8-7.0	Spinach	6.0-7.0
Cauliflower	6.0-7.0	Squash	5.5-6.5
Celery	6.0-7.0	Sweet corn	5.5-7.5
Chilli pepper	5.0-6.0	Sweet Potatoes	5.0-5.9
Cucumber	5.5-6.8	Tomato	5.5-6.8
Egg plant/ Brinjal	5.5-6.5	Turnip	5.0-7.0
Lettuce	6.0-7.0	Watermelon	5.5-6.5
Onion	6.0-6.7	-	-

Source: Havlin *et al.,* 2005 and Splittstoesser, 1990)

Table 14 : Relative tolerance of vegetable crops to soil acidity

Slightly tolerant (pH 6.0-6.8)	Moderately tolerant (pH 5.5-6.8)	Highly tolerant (pH 5.0-6.8)
Beat	French bean	Pumpkin
Broccoli	Carrot	Sweet Potato
Cabbage	Cucumber	Chow-chow
Celary	Brinjal	Yam
Spinach	Knolkhol	Dioscoriaspp
Chinise cabbage	Pea	
Leek	Chilli	
	Pumpkin	

Soluble Salts

The term soluble salts refer to the inorganic soil constituents that are dissolved in the soil later. The measurement of the electrical conductivity of a soil extract gives an indication of the total concentration of salts in the soil. In saline soil the salt concentration in the soil solution is very high and as a result the osmotic pressure of the soil solution is also very high. Crops differ in their sensitivity to soluble salts. High soluble salts can restrict root growth, cause burning of the foliage, and limit the crop yields. Most soils in western and southern regions are generally non-saline (0 to 2 ds/cm) but a few may be high in soluble salts. Sometimes, when fertilizers are over applied or placed too close to the roots the soluble salts may limit the plant growth. Relative salt tolerance of various fruit and vegetable crops is presented in Table 15.

Table 15 : Soluble salt test values and relative salt tolerance of fruits and vegetables

Non-tolerant (0-2 dSm^{-1})	Slightly-tolerant (3-4 dSm^{-1})	Moderately tolerant (5-7 dSm^{-1})	Tolerant (8-16 dSm^{-1})
Carrot	Apple Cabbage	Table beats	Asparagus Swiss chard
Green Bean	Grapes Lettuce	Cucumber Muskmelons	
Onion	Peppers Potatoes	Squash Tomatoes	
Radish	Sweet Corn	Spinach	

Soil pH modification using amendments

Liming materials are used to increase the pH of soils. Not all soils need to be limed. Many soils in eastern and southern region were formed under conditions of higher rainfall and tend to be acidic. Intensive cropping and continuous use of manures and/or ammonium-based ferilizers have a potential to acidify soils over time. In contrast, if high pH irrigation water is used on acid soils, soil pH may actually increase over time. The need for lime is determined from a routine soil test. The pH value reported in a soil test indicates whether lime is needed. In most cases, the quantity of lime required is related to soil texture and organic matter content. At the same soil pH, soils with a high clay and organic matter content will require more lime than the soils with high sand and low organic matter content. Of all calcium compounds, gypsum is considered to be the best and cheapest amendment for the reclamation of sodic soils.

Use of manures, fertilizers and bio-fertilizers

Fertilizers supply plant food and help to increase the yield of different crops through the improvement in the soil fertility. Manures not only supply plant nutrients but also improve the soil physical environment influencing plant growth. Bio-ferilizers are about to make a significant contribution towards the development of strategies for productivity improvement which do not lead to an exponential rise in the consumption of nonrenewable forms of energy. Plant roots absorb the nutrients from the soil solution in the ionic (inorganic charged) form. Larger molecules can also be absorbed by roots, but their rate of absorption is slow. Thus, if a fertilizer (organic or inorganic) is applied, it must first be broken down to its simplest forms to be used efficiently by plants. The amount of plant nutrients in a fertilizer programme can be reduced if organic manures are used. The nutrient content of manures varies with type of livestock and methods used in storage, handling, and application. As a general rule, the suggested rates of N, P_2O_5 and K_2O can be reduced by 5.5, 2.2, and 5.5 kg for each tonne of wet manure applied per hectare respectively. Therefore, it is essential to evolve and adopt a strategy of integrated nutrient supply by using a judicious mix of chemical fertilizers, organic manures and bio-fertilizers in relation to their importance, possible reactions in soils etc. The nutrient contents of common organic manures and inorganic fertilizers are given in Table 16.

Table 16 : Nutrient content of common organic manures and inorganic fertilizers

Name	Nutrient content (%)		
	N	P_2O_5	K_2O
Fertilizers			
Ammonium Sulphate	20.5	-	-
Ammonium Chloride	28.0	-	-
Ammonium Nitrate	32.5	-	-
Calcium Ammonium Nitrate (CAN)	26.0	-	-
Calcium Nitrate	13.0	-	-
Urea	46.0	-	-
Single Super Phosphate (SSP)	-	16.0	-
Diammonium Phosphate	18.0	54	-
Rock Phosphate	-	20.0-23.0	-
Basic Slug	-	6.0-8.0	-
Potassium Chloride (MOP)	-	-	60.0
Potassium Sulphate	-	-	48.0
Potassium Nitrate	13.0	-	37.0
Organic Manures			
FYM	0.6-2.1	0.26-1.10	1.0-3.6
Poultry Manure	2.0-4.5	4.5-6.0	2.1-2.4
Compost	1.5-3.5	0.5-1.0	1.0-2.0
Vermi-compost	1.21	0.86	1.01
Biogas Slury	1.06	0.52	0.70
Ground nut Cake	7.30	1.50	1.30
Sesamum cake	6.20	2.00	1.20
Neem cake	5.20	1.10	1.50
Castor cake	4.40	1.10	1.40
Coconut cake	3.00	1.90	1.80
Cotton seed cake	4.00	1.90	1.60
Sewage Sludge	1.0-6.0	0.75-2.0	0.40
Dhaincha	3.50	0.60	1.20
Sesbania	2.80	0.10	1.40
Sunhemp	2.30	0.50	1.80
Composted Coir pith	1.06	0.40	1.20
Fish Meal	4.00	3.90	1.80
Blood meal	10.10	1.20	1.00
Goat Manure	2.40	0.90	2.00
Sheep Manure	1.93	1.70	2.30
pig manure	3.70	3.30	0.40
Night Soil compost	1.40	2.80	4.00

Use of cheap sources of nutrients Cost of nutrients is a major expenditure in crop production. As far as possible attempts must be made to reduce the cost on nutrients through use of cheep sources of nutrients like rock phosphate, gypsum, pyrites, crop residues etc.

Rock phosphate: In acid soils rock phosphate can be used directly as a source of phosphorus. It is more convenient to use rock phosphate in perennial crops than annual crops as the P from rock phosphate becomes slowly available.

Gypsum: It is the cheapest source of sulphur available today. Gypsum as a source of S can be used on any type of soils. It will give additional benefits as a source of Ca in acid soils.

Pyrite: Pyrites can be used as a source of both sulphur and iron in neutral and alkaline soils. Use of pyrite in alkaline soils helps in increasing the availability of other nutrients by reducing the soil pH.

Bio-fertilizers: Bio-fertilizers are inputs containing useful microorganisms. They are capable of mobilizing plant nutrients from unavailable forms to available forms through biological processes. They are cost-effective and inexpensive sources of plant nutrients. Bio-fertilizers are grouped under three distinct categories viz., N-fixers, phosphate solubilisers and mycorrhizal fungi. Nitrogen-fixing organisms associated with horticultural crops are *Rhizobium* sp. Associated with legume vegetables and free - living microorganisms belonging to *Azotobacter* sp. and *Azospirillum* sp. which live in association with root systems of non-leguminous horticultural crops. Several soil bacteria particularly those belonging to the genera *Pseudomonas* and *Bacillus* and fungi belonging to the genera *Penicillium* and *Aspergillus* posess the ability to convert insoluble P in soil into soluble forms.

The vesicular rbuscularmycorrhizae (VAM) are formed by the fungi belonging to the genera *Glomus*, *Gigaspora*, *Acaulospora* and *Sclerocystis* which produce vesicles system. Arbuscles are highly branched fungal hyphae and the vesicles are bulbus swellings of these fungal hyphae. These fungi make nutrients available to the host plants. Chemical fertilizers and bio-fertilizers should not be applied together as the microorganisms in bio-fertilizers may get killed.

Composts from weeds and other bio wastes: Large quantity of weed biomass is available in rural areas during monsoon season which are rich sources of plant nutrients. They can be both directly incorporated into the soil before seeding or can be converted into enriched compost and used in crop production.

Importance of balanced fertilization in horticultural crops

Conceptually integrated nutrient management would essentially mean rational use of fertilizers and organic manures for supply of nutrients for production in such a manner that would ensure (i) maximum efficiency of fertilizer use, (ii) harnessing best possible positive and synergistic interactions amongst the various factors of production (planting material, water, agrochemicals etc.,), (iii) least adverse effect on environment by minimizing nutrient losses, (iv) maintaining soil productivity, and (v) sustaining high yields commensurate with the biological potential of the crop variety under the unique soil, climate and agro ecological set-up. Integrated nutrient management for a crops/cropping systems is the only viable strategy advocating accelerated and enhanced use of fertilizers with matching adoption of organic manure and bio-fertilizers so that productivity is maintained for a sustainable agriculture. Balance has to be maintained in the soil-crop system and it has also to take care of all other factors of production and make allowances for residual effects of past fertilizer applications, N fixation etc., and to ensure that there is no toxicity/deficiency of any element. Perennial fruit crops are quite a bit different from annual crops. Annual crop growers can almost predict what their final yield will be from the amount of fertilizers they apply. In case of perennial fruit crops more than 90% of the absorbed nutrients get locked-up in the woody part of the tree and the current application will not influence the yields of current season. It is because of this, yield predictions to applied nutrients are difficult in perennial fruit crops. However, the deep root systems help them to bring in nutrients from a large volume of soil and can store nutrients in the vegetative part and resume them again and again during later growing seasons. Thus the annual crops (vegetables, flowers, etc.) and perennial crops need to be tackled differently under integrated nutrient management strategies. Because of this complex nature of perennial fruit crops growers seem to worry most about, whether they are applying enough fertilizers. How can we balance nutrient applications under these complex situations? Growers believe that if they spend money to add fertilizers to spray some foliar nutrient mixtures then their yields would increase. However, an immediate benefit in such case is not seen. This also leads to imbalanced use of nutrients and leads to many complications in fruit production. Too much nitrogen fertilisers result in excess vegetative growth and reduce the reproductive growth which otherwise is needed to maintain large yields of fruits. Excess nitrogen can result in soft fruits with poor colour in apple and predispose vegetables to attacks of pests and diseases. It will delay ripening in grapes. A thick vegetative canopy increases diseases by slowing the drying. Too much potassium can cause bitter pit in apples and mango by upsetting calcium uptake. The ultimate effects of these are that yields do not increase. Hence in crop production one should know the concentration of nutrients in plant tissue before fertilising the trees and for mid-term corrections in vegetables.

Methods to enhance nutrient use efficiency source: Commonly used prilled urea is modified to improve its use efficiency by making it slowly soluble through physical transformation (eg. Urea super granules (USG), coated with lac etc.,) and chemical/biological modifications (sulphur-coated, neem oil/extract-coated, guar-gum-blended, tar-coated etc.). Urea is also useful for foliar application and fertigation. Ammonium sulphate is not recommended in high rainfall areas due to its high residual acidity. Calcium ammonium nitrate is useful where supplement with calcium is necessary. Single superphosphate and DAP are the most common phosphatic fertilizers used. Muriate of potash is widely used potassic ferilizer. In crops like potato and grape in which chlorides interfere with quality, sulphate of potash is recommended. The latter is also convenient for foliar application and fertigation. Several compound/complex fertilizers are available that contain more than one major nutrient but need to supplement with straight fertilizers to meet the exact needs of the crops.

Time and method of application

The key to enhance ferilizer use efficiency is to synchronize the time of ferilizer application with the growth need of the crop and periodof high root activity. It is useful to increase the number of split applications provided the cost of application is not prohibitive. In fruit crops, it is best to apply fertilizers prior to vegetative flushing, usually as pre- and post monsoon applications. In vegetable crops it is best to complete application of all the splits before the grand growth / flowering stage. Although the same is recommended for potassic fertilizers, split applications are found to be beneficial in banana and some long duration vegetable crops like chilli, capsicum and tomato. It is best to apply fertilizers as a band close to the zone of high root activity followed by proper incorporation into soil. In most vegetable crops application of fertilizers at 5 cm below the seed/ plant row ensures maximum utilisation of the applied fertilizer and may economize on N and P input in some crops. In fruit crops, placement of fertilizers in a band of 15-90 cm width midway between the trunk and drip circle ensures application close to the zone of high root activity and thereby enhanced utilization of the fertilizer nutrient.

Foliar fertilization

An appropriate time to consider foliar fertilization would be when a shortage of a nutrient is evident as indicated by tissue analysis or visual symptoms. In these situations, foliar fertilization provides the quickest means to correct the problem. Certain soil conditions, such as high pH, excess moisture, or cool temperatures, may render a nutrient or nutrients unavailable to the plant root. If these conditions exist, the problem may be more effectively corrected by foliar applications compared with soil applications. Two to three sprays of micronutrients are needed to meet crop demands. Nutrients applied to the leaves can be absorbed and utilized by the plant. However, for nitrogen, phosphorus, and potassium the quantity absorbed at any one time is small relative to the larger levels required for growth by the plant foliar application of these three nutrients cannot be expected to supply the total amount required for crop production.

Drip irrigation and fertigation

Drip or trickle irrigation is a type of irrigation where water is supplied under low pressures directly to or near the plant's root zone. Water is carried through plastic tubing and emitted through small openings. Drip irrigation is often used in combination with plastic mulch. Advantages of using drip irrigation include better control of foliar diseases and more efficient water and ferilizer use. Water savings with drip irrigation can amount to as much as 50% compared with overhead sprinkler system. This method of irrigation is particularly suited for high value crops such as tomatoes, peppers, vine crops, and cole crops. Fertigation refers to the application of water soluble fertilizer through the irrigation water. Nutrients in a concentrated solution are injected in the irrigation water using an appropriate injection device. Providing nutrients through the irrigation system enables more flexibility in a fertilizer programme. The type of system selected will depend on the crop being grown and resources available.

Nitrogen is the primary nutrient applied through the system. Urea is the most economical source of nitrogen to apply; potassium nitrate and ammonium sulphate are soluble and can also be used. Calcium nitrate is also water-soluble but may precipitate if injected in high pH water. Drip irrigation, in combination with plastic mulch, allows for precise timing (spoon feeding) of nitrogen. Small amounts can be applied daily (0.5-1.0 kg N/ a) or weekly (2.5-5.0 kg N/a) to meet the growth demands of the crop. Potassium can also be injected without any precipitation problems, although in most soils, a broadcast and starter application can meet plant requirements. Phosphorus may precipitate with micronutrients or with calcium and magnesium in the irrigation water resulting in clogging problems. Some micronutrients such as copper, iron, manganese, and zinc may also precipitate in high pH water. For most situations, P and micronutrients, if needed, should be applied to soil before planting. These elements can be injected alone in the drip system without precipitation problems. For P application through drip system phosphoric acid should be used. For micronutrients, chelated forms should be applied to soils. Clogging problems in drip lines can be corrected by injecting acids into the line to dissolve precipitates. If bacteria or algal growth causes clogging, then chlorine should be mixed with water. In all cases, the cause of clogging should be determined before treatment, and injection rates of chlorine or acid should be carefully monitored to avoid damage to the system and the plants.

Assessing fertilizer recovery

For getting precise information on nutrients removal by the crops, fertilizer use efficiency and nutrient recovery. The total analysis of whole plant is must and demand of the day.

Biological parameters: Nutrient recovery and fertilizer use efficiency are agronomical parameters which help in understanding the performance of fertilizers/manures applied to the crop as well as optimum nutrient dose needed to raise the crop. For this, whole plant is taken from the check plots (C) and fertilized plots (T) during growth or at harvest. Samples are washed dried and analyzed for nutrients. The nutrient element concentration is multiplied by the dry matter yield and nutrient uptake pertaining to the element is worked out. Other biological and agronomical parameters are calculated as under:

i) $$\text{Nutrient requirement (NR)} = \frac{\text{Nutrient uptake (kg/ha)}}{\text{Yield (q/ha)}}$$

ii) $$\%\ \text{Nutrient recovery} = \frac{\text{Nutrient uptake}_{\text{Treated plot}} - \text{Nutrient uptake}_{\text{Control plot}}}{\text{Fertilizer nutrient rate } (\text{kg / ha})} \times 100$$

iii) $$\%\ \text{Nutrient utilization efficiency} = \frac{\text{Yield}_T - \text{Yield}_C}{\text{Nutrient uptake}_T - \text{Nutrient uptake}_C} \times 100$$
$(\text{q yield / kg nutrient absorbed})$

iv) $$\%\ \text{Soil efficiency } (\text{SE}) = \frac{\text{Nutrient uptake}(\text{kg / ha})}{\text{Soil test value } (\text{kg / ha})} \times 100$$

v) $$\%\ \text{Fertilizer efficiency } (\text{FE}) = \frac{\text{Nutrient uptake} - \text{Soil test value X SE}}{\text{Fertilizer nutrient rate } (\text{kg / ha})} \times 100$$

vi) $$\text{Fertilizer dose } (\text{kg / ha}) = \frac{\text{NR}}{\text{FE} \times 100} - \frac{\text{SE}}{\text{FE}} \times \text{Soil test value } (\text{kg / ha})$$

Radio assay parameters: Apart from the biological and agronomical parameters defined above, the following radio assay values are helpful in working out the nutrient use efficiency in the soil / plant system with help of radio isotope.

%dff: The amount of nutrient in the plant derived from applied carrier expressed as % dff was worked out from the activity data without assessing the yield.

$$\%\ \text{dff} = \frac{\text{Specific activity of sample}}{\text{Specific activity of carrier}} \times 100$$

% nutrient utilization: Because of other means, it is not possible to distinguish the nutrient removal by the plant from soil or carrier; isotopes provide the only direct method of quantitatively measuring the efficiency of the plant nutrient use efficiency of fertilizer

$$\% \text{ Utilization } = \frac{\% \text{ dff x Yield of nutrient in plant (ppm)}}{\text{Rate of fertilizer nutrient applied (ppm)}} \times 100$$

A value: The "A" value concept is based on the assumption that when two sources of same nutrient are present in the soil, the plant will absorb nutrient from each source in proportion to their respective quantities available. The A values and % dff are independent of crop yield; the potential error in determining the biological yields are thus avoided.

$$\text{A value} = \frac{\% \text{ dfs}}{\% \text{ dff}} \times \text{Rate of fertilizer nutrient applied (kg / ha)}$$

Fertilizer recommendations based on soil test

Soil test reports form a basis of sound recommendations on plant nutrient application through fertilizers and /or manures. The soil test based fertilizer use enhances productivity, improves nutrients efficiency and minimizes the risk environment pollution. Fertilizer rates needed for a crop generally depends on indigenous nutrient supply and very with the nature of crop, its yield potential, previous crop grown and agronomic management practices. Balance nutrition of crop should ensure that the nutrients in available forms are in adequate quantities and right proportions as per the requirement of the crops. Therefore, the addition of fertilizers and manures should be such that in combination with indigenous nutrient supply, they provide the nutrients in appropriate quantities and ratios matching the plant requirement for desired yield goals. Different approaches of fertilizer recommendations in vogue are briefly discussed here (Singh *et al.,* 2013).

Generalized fertilizer recommendations (or) Agronomic fertilizer recommendations

These recommendations are based on multilocational trials conducted with graded doses of N, P and K fertilizers (Table. 17). Yield responses to applied nutrients are worked out and an optimum dose of fertilizers is recommended for individual crops in a given agro-ecological region. The major drawback with this approach is that the variation in soil fertility is not taken into consideration. Obviously under high or low fertility conditions, the applied nutrients prove to be either wasteful or remain insufficient. Hence the use efficiency of fertilizer nutrients often remains low with the adoption of these recommendations. Reports from long-term fertilizer experiments revealed a considerable built-up in soil P status due to continuous application of general recommended dose in intensive cropping on alluvial, black and red soils. In certain cases the built-up of available P reached a level where no more phosphate application is needed for next few seasons

(Nambiar, 1994; Singh *et al.,* 1998). At several other locations, even recommended rates of NPK proved sub-optimal when nutrient turn-over in soil-plant-system was higher under cereal based systems (Swaroop and Wanjari 2000).

Use of K has been quite low as compared to removal by the crops, leading to negative K balance and mining from the soil (Tiwari, 2002). Despite these limitations, general recommendations are being used extensively in India and many other Asian countries for their simplicity and exclusion of costs involved in soil testing, As the present day globally competitive agriculture necessities for an enhancement in input use efficiency (Singh *et al.,* 2013).

Table 17 : Generalized fertilizer recommendations of arid vegetable

Crop	Varieties	Dose in kg/ha		
		N	P_2O_5	K_2O
Clusterbean	Thar Bhadvi	60	60	60
Kachari	AHK-119	80	40	40
Kakadia	AHS-82	80	40	40
Beans	Thar Maghi	80	40	40
Water melon	Thar Manak	80	40	40
Bottle gourd	Thar Samridhi	80	40	40

Recommendations based on soil fertility rating

Most of the recommendations prescribed by the soil testing laboratories in India are based on soil fertility ratings (Table 18), the medium soil fertility being equated with general recommended dose. In low and very low or high and very high categories the fertilizer doses are raised or lowered by 25 to 50 percent of the general recommended dose as per situation. Since, nutrient recommendations with this approach are irrespective of yield goal and variation in soil type. These may perform well for moderate yield levels under optimal soil conditions. For high yield expectations as under commercial farming, however the approaches taking into account yield target are more suitable.

Table 18 : Recommendations based on soil fertility rating

Nutrient	Low*	Medium**	High***
N	<280	280-560	>560
P (Olsen)P (Bray)	<10<35	10-2535-80	>25>80
K	<120	120-280	>280

*In low and very low categories, the fertilizer dose is raised by 25 to 50% of the general recommendation. **In medium categories, the general recommendation is applicable. ***Whereas, for high and very high categories, the fertilizer dose are lowered by 25 to 50% of the general recommendation.

Fertilizer recommendations based on soil critical levels

The critical level concept was developed by Cate and Nelson (1965). Critical limits is the level of soil available nutrient above which sufficiency level sets-in, and that nutrient remains no longer a primary limiting factor from crop production (Table 19). This concept considers that soil test values should be inversely proportional to the yield gains from applied nutrients. Thus, probability of getting economic response to fertilizer application in the soil below the critical limit is quite high, whereas in soils below the critical limit is quite high, whereas in soils having the nutrient above the critical limit, such a probability is quite low. The critical level varies depending on the soil types, crops and varieties, soil test methods used and seasonal variations. This concept is useful for categorizing the soil into responsive and non-responsive groups, but the quantification of fertilizer dose for individual situations is not possible. It is more suitable for micronutrient recommendations for responsive category soils (Table 11&12).

Table 19 : Fertilizer recommendations based on soil critical levels

Nutrient	Limit (ppm)
Fe	4.5
Mn	1.0
Zn	0.6
Cu	0.2
B	0.5
Mo	0.1

Recommendations based on targeted yield

A significant linear relationship has been reported between yield of crops and nutrient uptake. This forms the basis for targeted yield concept initially advocated by Troug (1960). Rammanmoorthy *et al.* (1967) put forth the concept by which fertilizer recommendations for high yielding varieties for crops could be made. The needed parameters are

i) Nutrient requirement in kg/Quantal of production.

ii) Per cent contribution from soil available nutrients (CS).

iii) Per cent contribution from fertilizer nutrients (CF).

Such basic data on number of crops / varieties have been generated under the AICRP for investigations on Soil Test Crop Response Correlation for different soil climatic conditions for the purpose of making fertilizer recommendations.

The recommendations based on this approach are more quantitative, precise and meaningful because these involve both soil and water analysis and hence provide a scientific basis for balanced fertilization. Advantages of this approaches is that a farmer can select a suitable target according to his source and management conditions. The experience gained from the experimental farms and farmers field indicate that with this approach the yield target can be achieved with ± 10 percent deviation. These recommendations have the potential to meet the needs of the soil and crop production strategies such as :

1. Efficient use of fertilizers according to soil fertility status and crop species.
2. Choice of fertilizer recommendations under the fertilizer credit shortage.
3. Maintenance of soil fertility in cropping systems,
4. Identifications of suitable crop rotations from the point of view of relative ability of crops and crop varieties to utilize soil /fertilizer nutrients and
5. Area wise fertilizer recommendations based on yield targeting and soil nutrient index. Sharma *et al.* (1999) evolved following basic data and common equations for high yielding varieties of potato in different location of potato growing areas.

Artificial gradient concept: A site specific targeted yield concept consists of preparing artificial fertility strips and later laying out field trials with selected manuring treatments (22 for the plains and 14 for the hills) of N, P and K (Ramamoorthy, 1975). Linear models for prescription of the optimal fertilizer dose taking into account the soil test values have been developed (Table 20).

Table 20 : Targeted yield equations for potato production

Location	Soil type	Variety	Prescription equation	Reference
Ludhiana (Punjab)	Alluvial	Kufri Chandramukhi	FN = 0.88T-0.33SNF P_2O_5 = 0.71-0.875 SPFK_2O = 0.76 T-0.32SK	Valayutham (1979)
Jaladhar (Punjab)	Alluvial	-	FN = 1.21T-0.76 SNFP_2O_5 = 0.44T-2.93SPFK_2O + 1.36T-1.50SK	Singh (2005)
Pantnagar (Uttrkhand)	Mollisol	Kufri Chandramukhi	F = 1.19T-0.98SNF P_2O_5 = 0.50T-0.3.3 3SPFK_2O = 0.47T-0.60SK	Adinarayana (1987)
		Kufri Dewa	FN = 1.09 T-0.90S NF P_2O_5 = 0.70T-4.47 SPFK_2O = 0.38T-0.41SK	Adinarayana *et al.* (1984)
Ranchi (Jharkhand)	Calcareous	Kufri Sindhuri	FN = 1.50T-0.24SNF P_2O_5 = 1.35T-4.26 SPFK_2O = 0.81T–0.57SK	Prasad *et al.* (1987)

T = Targeted yield, SN, SP and SK are the soil test values and F is fertilizer dose

At Jaladhar, this has helped in increased fertilizer use efficiency and benefit: cost: ratio by 109 and 3.3%, respectively over the recommended fertilization without affecting the tuber yield. Moreover input cost on fertilizers decreased by over 50% and net returns increased by 5% when compared with recommended dose (Singh, 2005). Upadhayay and Sharma (2002) developed variety specific fertilizer prescription equations for judicious use of N, P and K for the western Indo-Gangetic plains and subsequently verified them on the cultivators' fields in Meerut area of UP for four commercial potato cultivators (Table 21).

This technique some times fails as results emanating may not represent true picture than those obtained from native soil fertility and also prime effect of nutrients contribution from fertilizers to the soil not taken in to consideration.

Native fertility concept: Another technique utilizes soil variation between, as well as within blocks by selecting fields differing in native fertility. Here the fields differing in native fertility are selected and an experiment in a 4^3 confouded design is laid out with various levels of N, P and K (Sharma and Sharma, 1987).

Table 21 : Targeted yield equations for predicting P and K dose for potato cvs.

Variety	Nutrient	Regression equation
Kufri Bahar	Nitrogen	FN = 2.67T-2.265SN
Kufri Badshah	-do-	FN = 2.48T-0.784SN
Kufri Jyoti	-do-	FN = 2.86T-2.161SN
Kufri Sutlej	-do-	FN = 1.45T-2.150SN
Kufri Bahar	Phosphorus	FP_2O_5 = 2.685T-13.285SP
Kufri Badshah	-do-	FP_2O_5 = 2.634T-12.316SP
Kufri Jyoti	-do-	FP_2O_5 = 0.892T-4.212SP
Kufri Sutlej	-do-	FP_2O_5 = 1.725T-10.414SP
Kufri Bahar	Potassium	FK_2O = 3.249T-4.70SK
Kufri Badshah	-do-	FK_2O = 2.429T-3.12SK
Kufri Jyoti	-do-	FK_2O = 2.910T-1.02SK
Kufri Sutlej	-do-	FK_2O = 2.595T-3.42SK

SN, SP and SK indicate available N, P and K

It was observed that $NO_3 - N$, organic carbon and available K accounted for 59, 25 and 41% variation in yield. In presence of N, contribution from soil P increased by 6.8% and in presence of N and P, contribution of K increased from 35.5 to 52.5%. The fertilizer adjustment equations (Table 22) for Shimla showed that fertilizer N dose was dependent on soil N and varied between 118 to 152 kg N/ha in soils containing $NO_3 - N$ in range of 95-124 ppm.

On alluvial soils of Jalandhar, N needs for potato varied from 150 to 195 kg N/ha as $K_2Cr_2O_7$- oxidisable N decreased from 480 to 288 ppm (Singh and Sharma,

1982; Singh and Grewal, 1985).The results showed that soil tests are a valuable tool in selecting fertilizer dose as per yield level thereby saving on fertilizers without affecting soil health.

Table 22 : Fertilizer adjustment equations for optimum potato yields

Year	Fertilizer adjustment equations	Fertilizer dose (kg/ha) N	K	Tuber yield (t/ha)a	Validity rangae
1981	FN=179.6-1.22SN-0.51SK-142.92r	128.5	-	30.8	95.38<SN<123.50 and 122.94<SK<264.6
1982	FN=198.3-31.21 SOC -172.42r	129.9	-	26.7	1.50<SOC<2.48
	FK=257.1-61.8SOC-178.67r	-	143.6	27.1	
1983	FN=151.3-0.135SN-54.82r	133	-	41.3	42<SN<147 and 84 SK<462

SN, SOC and SK indicate available N, soil organic matter and available K

Ready Reckoner of fertilizer doses at varying soil test values for specific yield target for clusterbean in the arid region

The fertilizer prescription equations are developed for different soils, different varieties and for different regions. They also include IPNS based adjustment equations using different type of local available organic sources. Substitution of soil test values in the prescription equations gives us that particular fertilizer nutrient recommendation which need to be converted in to fertilizer quantity to be applied (using conversion factor). In case of IPNS based equations, FYM composition and the rate are to be substituted in the place of FYM given in the fertilizer prescriptions. These fertilizer recommendations are grouped in two groups mainly food grain crops and horticultural crops. Ready reckoners for some of the crops are given in Table 23. STCR report can be consulted for a number of crops including, medicinal, forage, vegetables and horticulture crops.

In the equation of STCR, integrated plant nutrient supply (IPNS) system has also been taken into consideration. The nutrient supplied through FYM /compost was adjusted while supplying the inputs and thus farmers can save the money by saving of costly fertilizer inputs and increase his income.

These equations were validated in on-farm demonstrations and the results revealed that not only the yields could be achieved closed to the targets but the net profit was also greater compared to general recommended dose or farmers practice. Similar types of fertilizer adjustment equations for developing soil test based recommendations on fertilizer or fertilizer + manure application have been worked out for diverse agro-ecologies (Subha Rao and sri vastava 2001; Sharma and Singh 2000), from which ready rockoners for different soil test values can be prepared and made use of for fertilizer use.

Table 23: Ready Reckoner of fertilizer doses at varying soil test values for specific yield target for clusterbean

Soil available nutrient (kg/ha)			**Fertilizer nutrient required (kg/ha) for yield target of clusterbean**					
KMnO4-N	**Olsens'P**	**Amm.Ac.-K**	**15 q ha^{-1}**			**18 q ha^{-1}**		
			N	P_2O_5	K_2O	N	P_2O_5	K_2O
80	15	170	44	39	15	60	54	30
90	20	190	39	27	15	55	42	23
100	25	210	35	20	15	51	30	16
110	30	230	30	20	15	46	20	15
120	35	250	26	20	15	42	20	15
130	40	270	21	20	15	37	20	15
140	45	290	16	20	15	32	20	15
150	50	310	12	20	15	28	20	15
160	55	330	7	20	15	23	20	15

Soil testing laboratory : ARS, Bikaner
Soil : Sobhasar, Khiran, Gajnar, Bhamatsar series
Crop and variety : Guar (RGC-986)
Target range : 15-18 q ha^{-1}
Soil nitrogen range : 80-160 kg ha^{-1}
Soil phosphorus range : 15-55 kg ha^{-1}
Soil potassium range : 170-330 kg ha^{-1}

Fertilizer adjustment equation
FN = 5.38T – 0.46 SN
F P_2O_5 = 5.07 T – 2.46 SP_2O_5
FK_2O = 4.86 T – 0.34 SK_2O

Recommended based on nutrient index

Fertilizer recommendations as per this approach is based on summarizing the overall availability status of a particular nutrient in a given area into low, medium and high categories according to the method given by Parker *et al.* (1951). The nutrient index is calculated as follows.

Nutrient Index= (NLX1) + (NM X2) + NH X3)/NT

Where, NL, NM and NH are the number of soil samples falling in low, medium and high categories, respectively and NT is the total number of soil samples analyzed. If nutrient index is <1.5, the soil fertility status is low, a value from 1.5 to 2.5 indicated medium fertility and a nutrient index value >2.5 refers to high fertility status. Later it was felt that these limits attach undue weightage to medium class, hence the limits were modified by Ramanmoorthy and Bajaj (1969) as <1.67 for low, 1.67 to 2.33 for medium and >2.33 for high fertility status. In this approach the variation in individual field is not taken into consideration, Hence the fertilizer recommendations are a gross generalization for a situation. The approach also ignores the yield targets. Only ad-hoc adjustments of the general recommendations are possible in this case (Singh *et al.,* 2013).

Nutrient indexing approach (Ramamoorthy and Bajaj, 1969) was followed to evaluate the nutrient status of the kachri growing arid region of Rajasthan as in this strategy the variation in individual field is not taken into consideration but the fertilizer recommendations are made on the basis of overall nutritional status of the study area. As per nutrient index study of the kachri growing areas of Bikaner district, soils were deficient in available nitrogen (1.05) and phosphorus (1.66), whereas in case of potassium it was medium (2.05). It shows that soil has great demand for nitrogen and phosphorus followed by potassium. Potassium was in adequate amount in most of the soils; hence fertilizers recommendation should be applied on the basis of soil test (Fig. 15).

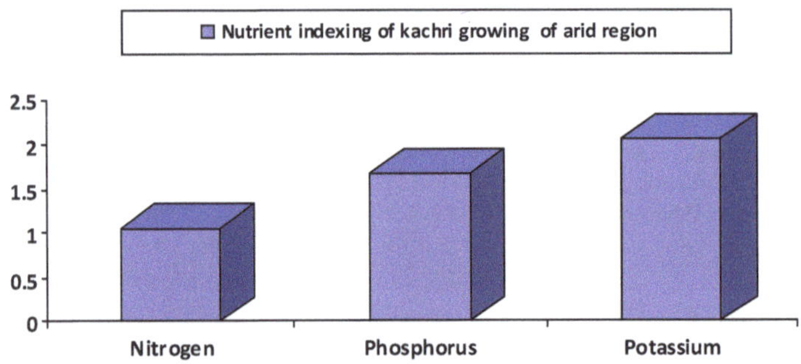

Fig. 15 : Nutrient indexing of the kachri growing of arid region of Bikaner district

Recommendation for cropping sequences

Meeting the fertilizer need of a cropping sequence as a whole is more important than those of individual crops. Legume-based cropping sequence can be taken up with lower fertilizer N doses due to the added advantage of symbiotic N-fixation. Also, utilization of phosphatic fertilizers by first crop of a sequence is very low (around 25%) and it is also not lost by any other process except erosion. A suitable adjustment in fertilizer recommendations can be made according to the crop needs, responses and residual effect in the cropping sequence. Use of FYM or green manure crop in the cropping sequence. Use of FYM or green manure crop in the cropping sequence also helps in increasing the availability of nutrients. A methodology for the prediction of post harvest soil test values developed by Sharma *et al.,* 1982 can be utilized for making fertilizer recommendations for the cropping system in the beginning itself, based on initial soil test (Singh *et al.,* 2005).

Use of simulation models

The conventional approaches of fertilizer recommendations based on crop responses and soil test values are often difficult to extrapolate to the areas/ regions differing in agro-climatic condition. In that case one option is to conduct extensive field experiments under individual recommendations. Alternatively, modern techniques of simulation modeling can be used for the purpose. These techniques are relatively cheaper, and take into account soil-plant-environment continuum. As these computer based models integrate information related to soil nutrient availability with other factors (e.g. management, climate etc.) affecting crop growth and response to applied nutrients, even conventionally developed fertilizer recommendations can be refined using the models. A number of models have been developed for the describing crop growth and nutrient uptake (Groot 1987; Grant, 1991; Singh, 1992). Crop simulation models developed by the International Fertilizer Development Centre (IFDC) in collaboration with International Benchmark sites network for Agro-Technology Transfer (IBSNAT) Project (Uehara 1989), and the simulation and systems analysis for Rice Production (SARP) project (Penning de Vries *et al.,* 1989) simulate crop growth and nutrient dyanamics and uptake process in responses to crop management and weather situation (Singh *et al.,* 2005).

Site -Specific nutrient management

Site-specific nutrient management (SSNM) strategies that include site and season-specific knowledge of crop nutrient requirements and indigenous nutrient supplies are required to increase productivity, yields, and nutrient use efficiency. The SSNM provides an approach for need based feeding of crops with nutrients while recognizing the inherent spatial variability. It involves monitoring of all pathways of plant nutrient flows / supply, and calls for judicious combination of fertilizers, bio fertilizers, organic manures, crop residues and nutrient efficient genotypes to sustain agricultural productivity. It avoids indiscriminate use of fertilizers and enables the farmer to dynamically adjust the fertilizer use to fill the deficit optimally between nutrient needs of the variety and nutrient supply from natural resources, organic sources, irrigation water etc. It aims at nutrient supply at optimal rates and times to achieve high yield and efficiency of nutrient use by the crop (Kumar *et al.*, 2016).

Application of different combinations of NPK doses significantly increased yield of *kachri* as compared to control. Application 120, 60 and 60 kg/ha of N, P_2O_5 and K_2O gave the highest *kachri* yield (108.87 q/ha) which was significantly higher than all other treatments, but at par with 80, 40 and 40 kg/ha of NPK and 80 and 40 kg/ha of NP. The increase in total yield was 7.85% higher over recommended NPK through fertilizers. Application of 100% NPK increased yield significantly by 81.63% compared to control. Whereas, this treatment gave only 2.10% more kachri yield as compared to recommended dose of fertilizers (Table 24).

Maximum yield was obtained when 120, 60 and 60 kg/ha of N, P_2O_5 and K_2O (108.87 q/ha) followed by 80, 40 and 40 kg/ha of N, P_2O_5 and K_2O (100.95 q/ha) and 80 and 40 kg/ha of N and P_2O_5 (98.87 q/ha) and 40, 20 and 20 kg/ha of N, P_2O_5 and K_2O (82.65 q/ha). Same trend was observed for percent yield response of different treatments.

Higher dry matter (%) was observed where 80 and 40 kg/ha of NK were applied followed by 120, 60 and 60 kg/ha of N, P_2O_5 and K_2O. However, dry matter yield (11.29 q/ha) was obtained where, 120, 60 and 60 kg/ha of N, P_2O_5 and K_2O followed by 80, 40 and 40 kg/ha of N, P_2O_5 and K_2O (Fig. 16).

Maximum per cent yield response was observed where 120, 60 and 60 kg/ha of N, P_2O_5 and K_2O was applied (95.88%) followed by 80, 40 and 40 kg/ha of N, P_2O_5 and K_2O (81.63%) and 80 and 40 kg/ha of N and P_2O_5 (77.89%) as compared to control. This might be due to more partitioning of dry matter to *kachri* as a result of balance nutrition in these treatment. (Fig. 17).

Table 24 : Requirements of N, P and K fertilizers of *kachri* using omission plot technique

Treatments	Yield (q/ha)	Average of fruit (g)
40, 20 and 20 kg/ha of NPK	82.65	37
80, 40 and 40 kg/ha of NPK	100.95	40
120, 60 and 60 kg/ha of NPK	108.87	42
40 and 40 kg/ha of PK	71.25	38
80 and 40 kg/ha of NK	80.45	40
80 and 40 kg/ha of NP	98.87	33
Without NPK (Absolute control)	55.58	29
CD at 5%	13.95	9

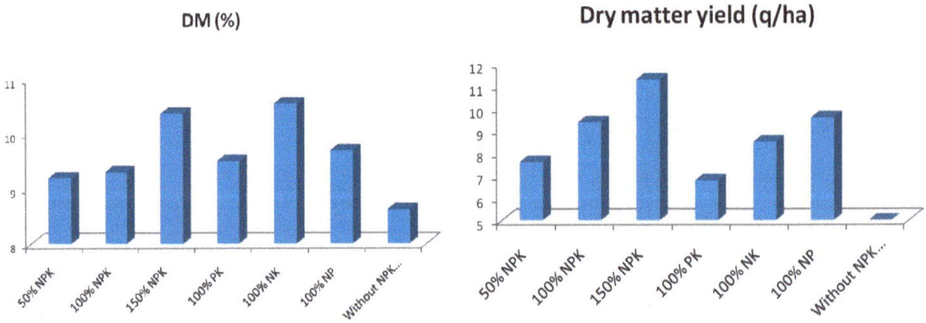

Fig 16: Dry matter percent and yield of *kachri* using omission plot technique

Net return and benefit cost ratio of the *kachri* from SSNM

Net return from *kachri* followed similar trend (Table-25) as that of fruit yield with highest values of Rs 178650/ha was observed in 120, 60 and 60 kg/ha of NPK treatment. This was closely followed by 80, 40 and 40 kg/ha of NPK (Rs 164827/ha) and 80 and 40 kg/ha of NP (Rs 161947/ha). 40, 20 and 20 kg/ha of NPK application gave net return of Rs 130263/ha as compared control (Rs 83708/ha). The benefit : cost ratio was highest in the treatment receiving 120, 60 and 60 kg/ha of NPK (4.57) closely followed by 80 and 40 kg/ha of NP treatments (4.52). Whereas, control gave the lowest B:C ratio (2.54) (Jatav *et al.,* 2016).

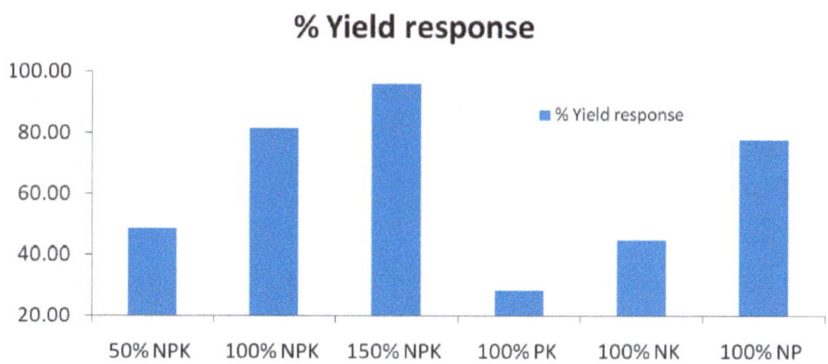

Fig 17: Percent yield response of *kachri* using omission plot technique

Table 25: Net return and B:C ratio under treatments

Treatments	Cost of cultivation	Yield q/ha	Price /kg	Gross return	Net return	B:C ratio
40, 20 and 20 kg/ha of NPK	35037	83	20	165300	130263	3.72
80, 40 and 40 kg/ha of NPK	37073	101	20	201900	164827	4.45
120, 60 and 60 kg/ha of NPK	39110	109	20	217740	178630	4.57
40 and 40 kg/ha of PK	36030	71	20	142500	106470	2.96
80 and 40 kg/ha of NK	35323	80	20	160900	125577	3.56
80 and 40 kg/ha of NP	35793	99	20	197740	161947	4.52
Without NPK (Absolute control)	33000	56	21	116708	83708	2.54

Nutrient requirement of cluster bean using omission plot technique to determine the requirements of P and K fertilizers

Application of 90 kg/ha each of P_2O_5 and K_2O gave higher grain yield of cluster bean followed by 60 kg/ha each of P_2O_5 and K_2O and without P +60 kg/ha K_2O as compared to other treatments (Table. 26 & 27). Same trend was observed for percent yield response of different treatments (Jatav *et al.,* 2016).

Table 26: Site specific nutrient management in cluster bean crop

Treatment	Vegetable cluster bean (q/ha)	Seed cluster bean (q/ha)	Straw yield (q/ha)
	27.48	6.74	77.12
60 kg/ha each of P_2O_5 and K_2O	41.25	9.06	114.45
90 kg/ha each of P_2O_5 and K_2O	43.45	11.25	112.45
Without P +60 kg/ha K_2O	35.36	6.48	67.48
Without K+60 kg/ha of P_2O_5	39.48	10.65	85.46
Without PK (Absolute control)	11.65	4.47	54.75
Mean	33.11	8.11	85.29

Table 27: Site specific nutrient management in cluster bean crop

Treatment	% Yield response in Vegetable cluster bean	% Yield response Seed cluster bean
30 kg/ha each of P_2O_5 and K_2O	135.9	50.8
60 kg/ha each of P_2O_5 and K_2O	254.1	102.7
90 kg/ha each of P_2O_5 and K_2O	273.0	151.7
Without P +60 kg/ha K_2O	203.5	45.0
Without K+60 kg/ha of P_2O_5	238.9	138.3
Mean	184.22	81.39

6

Water Quality Standards and Quality of Underground Irrigation Waters

Water quality standards (Classification)

Some countries have established national standards and criteria for evaluation and classification of water quality. The classification of irrigation waters has been made according to one or more of the following criteria.

1. Salinity or total concentration of soluble salts (EC)
2. Sodicity hazards (SAR)
3. Alkalinity hazards (RSC)
4. Combined evaluation on the basis of EC and SAR by USSL diagram
5. Combined evaluation on the basis of EC and Adj. SAR
6. Combined evaluation on the basis of EC, SAR and RSC

1. Salinity hazard (EC)

The classification of irrigation waters on the basis of EC is primarily based on the development of salinity in soils to extent that yields of crops are adversely affected. On the basis of electrical conductivity, the irrigation waters may be classified into six salinity classes as per table 28 (Gupta, 1986).

Table 28 : Salinity hazards and EC of Water

Salinity classes	EC of Water	Remarks
C_0- Non saline waters	EC < 0.2 dS m^{-1}	It may create severe permeability problem in the soils because infiltration rate into soil is adversely affected due to lack of salts in water to such an extent that the crop is not adequately supplied with water and yield is reduced.
C_1- Normal waters	EC 0.2–1.5 dS m^{-1}	Use for irrigation for most crops on most soils with little likelihood that soil salinity will develop.
C_2- Low salinity waters	EC1.5–3.0 dS m^{-1}	Use if a moderate amount of leaching occurs under the current irrigation practices. Most of the crops except sensitive ones can be grown on all soils

		except very heavy textured soils with impeded drainage
C_3-Medium salinity waters	EC 3.0 - 5.0 dS m^{-1}	Use on soils provided with good drainage such that leaching fraction is not less than 0.3, most of the semi tolerant and tolerant crops can be grown under good management. These waters may have moderate problem of high SAR and RSC.
C_4- High salinity waters	EC 5.0-10.0 dS m^{-1}	Suitable for irrigation on soils provided with very good drainage such that leaching fraction is always greater than 0.3, so that ECe does not exceed EC_{iw}. Only salt tolerant crops can be grown under good management practices. These waters may have the problem of high SAR but no problem due to RSC.
C_5- Very high salinity waters	EC > 10.0 dS m^{-1}	It is directly not suitable for irrigation but may be used in conjunction with low salinity water.

2. Sodium hazard (SAR)

The classification of irrigation waters with respect to sodic hazard on the basis of SAR is based primarily on the increase of exchangeable sodium and its effect on the physical conditions of soils. On the basis of SAR, irrigation water may be classified into six classes as per table 29 as proposed by Gupta (1986).

Table 29 : Sodium hazard (SAR)

Sodium hazards	SAR Levels	Remarks
S_0- Non sodic waters	SAR < 5	Use for irrigation on almost all soils for all crops even those sensitive to sodium
S_1- Normal waters	SAR 5-10	Use for irrigation on almost all soils with little danger of development of harmful levels of exchangeable sodium for growing all crops except sensitive to sodium.
S_2- Low sodic waters	SAR 10-20	Use for crops, which are semi tolerant or tolerant to sodium on almost all soils such that leaching fraction (LF) is around 0.3. If there is a presence of gypsum or calcium carbonate in soil, these waters can be used more successfully.
S_3- Medium sodic waters	SAR 20-30	Use only for crops which are tolerant to sodium on soils provided with good drainage such that leaching fraction is always greater than 0.3.
S_4- High sodic waters	SAR 30-40	This type of waters are directly not suitable for irrigation but may be used in cycle or conjunction with low sodicity waters or with the use of amendments such as gypsum.
S_5- Very high sodic waters	SAR >40	Waters are directly not suitable for irrigation without drastic treatment

3. Alkali hazard (RSC)

The carbonate or bicarbonate (alkali) hazard on the basis of RSC is primarily based on the precipitation of calcium and/or magnesium and pairing of residual carbonate (CO_3^-) or bicarbonate (HCO_3^-) with sodium and formation of sodium carbonate (Na_2CO_3) in the soil and increasing SAR/ESP characterizing it as alkali soil. RSC should be calculated for high pH (> 8.5) waters. On the basis of RSC, irrigation waters may be classified into six classes as per table 30 proposed by Gupta (1986).

Table 30 : Alkali hazard (RSC)

Alkali hazards	RSC	Remarks
A_0- Non-alkali waters	(RSC Negative)	Use for irrigation on almost all soils for all crops for indefinitely long periods without any problem.
A_1- Normal waters	RSC 0 me L^{-1}	Use for irrigation on almost all soils for all crops even those are sensitive to carbonates or bicarbonates
A_2- Low alkali waters	RSC <2.5 me L^{-1}	Use for irrigation on almost all soils for all crops.
A_3- Medium alkali waters	RSC 2.5-5.0 me L^{-1}	Use for irrigation on almost all soils with little danger of the development of harmful levels of alkali conditions for growing all crops except sensitive to carbonates or bicarbonates.
A_4- High alkali waters	RSC 5.0-10.0 me L^{-1}	It can be used for irrigation on soils provided with good drainage such that leaching fraction is not less than 0.3 for growing semi-tolerant and tolerant crops to sodium. EC should be < 3.0 dS m^{-1} and SAR should be <10.0
A_5- Very high alkali waters	RSC > 10.0 me L^{-1}	This water is directly not suitable for irrigation but may be used in conjunction with low alkalinity waters or with the use of amendments.

4. Combined evaluation on the basis of EC and SAR (by USSL diagram)

The USSL (1954) published a semi-logarithmic diagram, which rated waters with respect to salinity (EC) and sodium hazard (SAR) as shown in Table 31.

Table 31: EC and SAR values for different classes of salinity and sodium hazard of irrigation waters (Richards, 1954)

S.No.	Salinity hazard	EC (dS m^{-1})	Class	Sodium hazard	SAR	Class
1.	Low	0.10-0.25	C_1	Low	<10	S_1
2.	Medium	0.25-0.75	C_2	Medium	10-18	S_2
3.	High	0.75-2.25	C_3	High	18-26	S_3
4.	Very high	2.25-5.00	C_4	Very high	26-31	S_4

Each salinity class was associated with all the four sodium classes and vice-versa, thus, giving a total of sixteen classes for the purpose of evaluating the quality of water for irrigation. It is interesting to point out that out of sixteen classes, six classes (C_1S_1, C_1S_2, C_1S_3, C_2S_4, C_4S_1 and C_4S_2) were seldom found in nature (Bhumbla, 1977).

5. Combined evaluation on the basis of EC, adj. SAR and Boron concentration

Irrigation waters may be classified in five classes, each on the basis of adj. SAR (alkalinity) hazard, electrical conductivity (salinity) hazard and boron hazard. This classification may also be called as the ABC classification (Gupta, 1979) as presented in Table 32.

Table 32 : ABC classification of irrigation waters

S.No.	Adj. SAR	Class	Boron (mg L^{-1})	Class	EC (dS m^{-1})	Class
1.	<10	A_1	<3	B_1	<1.5	C_1
2.	10-20	A_2	3-4	B_2	1.5-3.0	C_2
3.	20-30	A_3	4-5	B_3	3.0-5.0	C_3
4.	30-40	A_4	5-10	B_4	5.0-10.0	C_4
5.	>40	A_5	>10	B_5	>10.0	C_5

Interpretations of quality class rating of the irrigation waters are as follows.

Table 33 : Adj. SAR

Quality	EC	Remarks
A_1-Normal waters	adj. SAR < 10	Use for irrigation on almost all soils without any problem in all crops.
A_2 – Low sodium waters	adj. SAR 10-20	Use for irrigation on light and medium textured soils without any problem and on heavy textured soils if well drained.
A_3–Medium sodium waters	adj. SAR 20-30	Use for irrigation on light textured soils but can not be used on heavy textured soils. The problem is alleviated if soils contain appreciable amount of native $CaCO_3$.

A_4– Sodium water	adj. SAR 30-40	Use on light textured soils for semi tolerant and tolerant crops, respectively. It can not be used for irrigation on heavy textured soils.
A_5 – High sodium waters	adj. SAR > 40	Water is not ideally suitable for irrigation and therefore, their use should be restricted as far as practicable.

Table 34 : Electrical conductivity

Category	EC	Remarks
C_1 – Normal waters	EC < 1.5 dS m^{-1}	Use for irrigation on most of soils for growing most of the crops.
C_2 – Low salinity waters	EC 1.5-3.0 dS m^{-1}	Water can be used for irrigation for growing most of the crops on light and medium textured soils but on heavy textured soils, only semi-tolerant crops could be grown.
C_3–Medium salinity waters	EC 3.0-5.0 dS m^{-1}	Water can be used for irrigation for growing semi-tolerant crops on light and medium textured soils and only tolerant crops on heavy textured soils.
C_4 – Saline waters	EC 5.0-10.0 dS m^{-1}	Water can be used for irrigation for growing tolerant crops only on light and medium textured soils. The soils have excellent drainage.
C_5 – High salinity waters	EC >10 dS m^{-1}	Water are not suitable for irrigation under ordinary conditions but may be used occasionally under very special circumstances.

Combined evaluation on the basis of EC, SAR and RSC

The guidelines based on EC, SAR and RSC for characterization of underground irrigation waters are suggested by Gupta *et al.* (1994) as presented in Table 35.

Table 35: Classification of underground irrigation waters based on EC, SAR and RSC

	Water quality	EC (dS m^{-1})	SAR	RSC (me L^{-1})
A.	Good	<2	<10	<2.5
B.	Saline			
i)	Marginally saline	2-4	<10	<2.5
ii)	Saline	>4	<10	<2.5
iii)	High-SAR saline	>4	>10	<2.5
C.	Alkali			
i)	Marginally alkali	<4	<10	2.5-4.0
ii)	Alkali	<4	<10	>4.0
iii)	Highly alkali	Variable	>10	>4.0

Salinity and sodicity indices of soils of Rajasthan

Accumulation of salts in soils takes place due to irrigation with saline water, evaporation and impeded internal drainage. Accumulation of salts in soil is dependent on soil type, temperature and salt composition of irrigation water. Muhr *et al.* (1965) and Seth (1967) classified the salt affected soils on the basis of salinity and alkalinity indices. The rating chart and limits for salinity and alkalinity indices are given in Table 36 and 37.

Table 36: Limits for salinity and alkalinity groups (Muhr *et al.,* 1965)

Groups	EC (dS m^{-1}) of 1:2 soil water suspension	Classification	pH (1:2 soil water suspension)	Classification
A	< 1.0	Normal	< 8.0	Normal
B	1.0-2.0	Tending to become saline (Critical for germination)	8.0-8.5	Alkaline
C	2.0-3.0	Saline (Critical for growth of salt sensitive crops)	8.5-9.0	Very alkaline (Soil tending to become alkaline)
D	> 3.0	Highly saline (Injurious for most of the crops)	> 9.0	Highly alkaline (Alkali soil)

Table 37: Rating chart of salinity and alkalinity indices (Seth, 1967)

Index	Rating for salinity	Rating for alkalinity
< 1.5	No salinity problem	No alkalinity problem
1.5-2.5	Mild salinity problem	Slight alkalinity problem
2.5-3.5	Moderate salinity problem	Moderate alkalinity problem
> 3.5	Severe salinity problem	Severe alkalinity problem

Sehgal *et al.* (1987) classified salt affected soils on the basis of salinity and alkalinity hazards. The limits for salinity and alkalinity groups are given in Table 37a.

Table 37a: Limits for salinity and alkalinity groups (Sehgal *et al.,* 1987)

Salinity classes			Alkalinity classes		
Groups	EC (dS m^{-1})	Classification	Groups	pH	Classification
Vs	0-2	Very slight	S	7.3-8.3	Slight to negligible
S	2-4	Slight	M	8.3-9.0	Moderate
M	4-8	Moderate	St	> 9.0	Strong
St	8-15	Strong			

Fertility status and indices of soils of Rajasthan

Rating and status of soils of Rajasthan

Subbiah and Asija (1956) and Muhr *et al.* (1965) classified the organic carbon, available nitrogen, available phosphorus and available potassium into low, medium and high classes as given below (Table 38):

Table 38: Rating chart for fertility groups

Nutrient	Low	Medium	High	References
Organic carbon (%)	< 0.5	0.5-0.75	> 0.75	Muhr *et al.* (1965)
Available nitrogen (kg N ha^{-1})	< 250	250-500	> 500	Subbiah and Asija (1956)
Available phosphorous (kg P_2O_5 ha^{-1})	< 20	20-50	> 50	Muhr *et al.* (1965)
Available potassium (kg K_2O ha^{-1})	< 125	125-300	> 300	Muhr *et al.* (1965)

The nutrient index below 1.5 is low, from 1.5 to 2.5 is medium and above 2.5 is high (Muhr *et al.*, 1965).

According to Department of Agriculture, Rajasthan, the soils of Rajasthan have been classified into seven fertility groups based on the analytical data. The names of the district along with the fertility group are given below (Table 39):

Table 39: Fertility grouping of rajasthan soils

S. No.	Name of the districts	Fertility group		
		Available N	Available P	Available K
1.	Ajmer, Alwar, Bikaner, Kota, Jodhpur, Jhunjhunu, Nagaur, Sirohi and Churu	Low	Medium	Medium
2.	Tonk, Banswara, Bilwara and Chittorgarh	Medium	Medium	Medium
3.	Bundi and Jaisalmer	Low	Low	Medium
4.	Barmer and Sriganganagar	Low	Medium	High
5.	Udaipur	Medium	Medium	High
6.	Dungarpur and Sikar	Low	High	Medium
7.	Sawai Madhopur	Low	High	Medium

On the basis of analysis, the fertility map of Rajasthan has been prepared. The fertility map indicates that out of 21 districts of Rajasthan, the soils in 16 districts were deficient in nitrogen while phosphorous was low only in two districts. The potassium status ranged from medium to high in Rajasthan.

Rating and status of micronutrients in soils of Rajasthan

The critical limits of available micronutrients are given in Table 40.

Table 40: Critical limits of DTPA extractable micronutrients (mg kg^{-1} soil)

Nutrient	Deficient	Marginal	Sufficient	Reference
Copper	< 0.2	-	> 0.2	Lindsay and Norvell (1978)
Manganese	< 2.0	-	> 2.0	Lindsay and Norvell (1978)
Zinc	< 0.5	0.5-1.0	> 1.0	Lindsay and Norvell (1978), Takkar and Randhawa (1978)
Iron	< 4.5	-	> 4.5	Takkar and Randhawa (1978), Lindsay Norvell (1978) and Dhir *et al.* (1983)

7

Plant Nutrient and Their Form

The essential plant nutrients, or essential elements, are listed in the Table 41. It enumerates 16 chemical elements of which the essentiality for plant growth and reproduction has been established according to the criteria formulated by D.I. Arnon and P.R. Stout, 1939. This listing has existed since 1954 with chlorine as the latest addition (cited by Hopkins 1999).

These plant nutrients are classified into various groups:

1. Major elements and trace elements : Of the 16 listed, 9 are major elements or macronutrients and 7 are trace elements or micronutrients. Major elements are those which are needed by plants for their growth and reproduction in relatively large amounts while trace elements are needed in smaller amounts compared to the former. The major elements are further sub classified into structural elements (carbon, oxygen and hydrogen), primary macronutrients (nitrogen, phosphorus and potassium), and secondary macronutrients (calcium, sulfur, and magnesium). Micronutrients are Zn, Cu, Mn, Fe, Mo_4 and boron (Table 41).

2. Mineral and non-mineral nutrients : Mineral nutrients are absorbed by plants primarily in ionic form from the soil while non-mineral nutrients are absorbed either from the soil or from the atmosphere as a component of compounds. The elements numbered 4 to 16 in the list are considered mineral nutrients. The carbon, oxygen, and hydrogen are considered non-mineral nutrients. They are sourced from carbon dioxide (CO_2) and water (H_2O) which are reactants in the process of photosynthesis.

Table 41 : List of essential elements or essential plant nutreints.

Element	Atomic weight	Forms absorbed	Range of concentrationin dry tissue*
Major Elements or Macronutrients			
1. Carbon (C)	12.01	CO_2	45% (450,000 ppm)
2. Oxygen (O)	16.00	CO_2, O_2	45% (450,000 ppm)
3. Hydrogen (H)	1.01	H_20	6% (60,000 ppm)
4. Nitrogen (N)	14.01	NO_3^-, NH_4^+	0.5-6% (5,000-60,000 ppm)
5. Phosphorus (P)	30.97	$H_2PO_4^-$, HPO_4^{2-}	0.15-0.5% (1,500-5,000 ppm)
6. Potassium (K)	39.10	K^+	0.8-8% (8,000-80,000 ppm)
7. Calcium (Ca)	40.08	Ca^{2+}	0.1-6% (1,000-60,000 ppm)
8. Sulfur (S)	32.06	SO_4^{2-}	0.1-1.5% (1,000-15,000 ppm)
9. Magnesium (Mg)	24.31	Mg^{2+}	0.05-1% (500-10,000 ppm)
Trace Elements or Micronutrients			
10. Iron (Fe)	55.85	Fe^{2+}, Fe^{3+}	20-600 ppm
11. Chlorine (Cl)	35.45	Cl^-	10-80,000 ppm
12. Manganese (Mn)	54.94	Mn^{2+}	10-600 ppm
13. Zinc (Zn)	65.38	Zn^{2+}	10-250 ppm
14. Copper (Cu)	63.54	Cu^+, Cu^{2+}	2-50 ppm
15. Boron (B)	10.81	BO_3^{3-}, H_3BO_3	0.2-800 ppm
16. Molybdenum (Mo)	95.94	MoO_4^{2-}	0.1-10 ppm

*Modified from Epstein (1994; ppm equivalent of concentration expressed in percent (%) is supplied).

The relative nutrient status of a plant may be easily described using the following terms as per given below:

Deficiency range	Nutrient deficiencies occur in plants when an essential element is not taken up by the plant in sufficient amounts. As a result, yield will be limited by the element which is deficient. While slight to moderate deficiencies do not always result in visual deficiency symptoms, distinct visual symptoms appear in severe cases.
Critical range	Below the critical range of nutrients, an addition of the essential element will trigger an increase in yield. Above the critical range, the levels of essential nutrients are considered sufficient.
Sufficiency range	Within this range, additions of the essential nutrient will not result in any increase in yield. However, uptake of the nutrient may continue. Thus, the concentration of that essential nutrient in plant tissue will also increase. We refer to the uptake of an essential nutrient within the sufficiency range as luxury consumption.
Toxicity range	Toxicities occur when an essential (or nonessential) element is taken up in great enough quantities to actually reduce plant growth. As a result, toxicities can severely limit yield.

8

Soil-Nutrient Relationships

Cation exchange: The 'soil cations' essential for plant growth include ammonium, calcium, magnesium, and potassium. There are three additional 'soil cations,' which are not essential plant elements but affect soil pH. The additional 'soil cations' include sodium, aluminum and hydrogen.

The major distinguishing characteristic of cations is their positive charge. Just like a magnet, a positive charge is strongly attracted to a negative charge. When soil particles have a negative charge, the particles attract and retain cations. These soils are said the have a cation exchange capacity. Although most soils are negatively charged and attract cations.

The 'soil cations' are further divided into two categories. Ammonium, calcium, magnesium, potassium, and sodium are known as the 'base cations,' while aluminum and hydrogen are known 'acid cations.' Unlike the other base cations, sodium is not an essential element for all plants. Soils that contain high levels of sodium can develop salinity and sodicity problems.

The words 'base' and 'acid' refer to the particular cation's influence on soil pH. As you might suspect, a soil with a lot of acid cations held by soil particles will have a low pH. In contrast, a highly alkaline soil predominately consists of base cations.

Cations in the soil compete with one another for a spot on the cation exchange capacity (Table 42 and Fig 18). However, some cations are attracted and held more strongly than other cations. In decreasing holding strength, the order with which cations are held by the soil particles follows: aluminum, hydrogen, calcium, potassium and nitrate, and sodium.

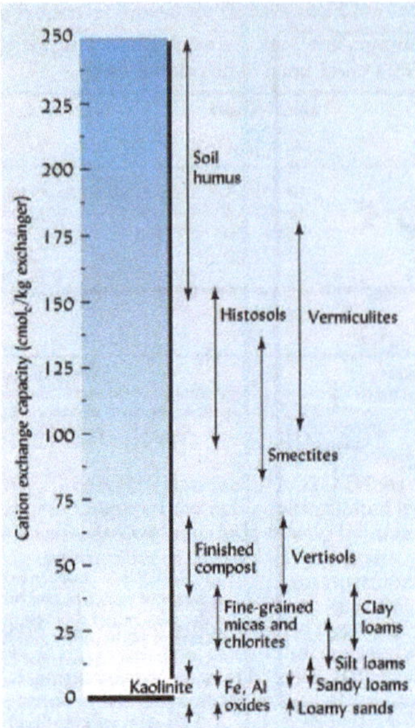

Fig. 18 : CEC values of various soil type, media, and minerals. Soils which have high amounts of organic matter and moderately weathered clays tend to have high CECs. As soils become highly weathered, the CEC of the soil decreases. Sandy soils, too, generally have lower CEC values. This is due to the lesser surface of sandy particles in comparison with clay minerals, which decreases the ability of sand particles to hold and retain nutrients.

Source: Brady and Weil. 2002. Elements of the Nature and Properties of Soil. Prentice Hall, New Jersey.

Table 42: Cation exchange data for representative mineral surface soils in different areas

Characteristics	Humid region soil (Alfisol)	Semi-arid region (Aridisol)	Arid region soil (Natragids)[a]
Exchangeable calcium (cmol/kg)	6-9	14-17	12-14
Other exchangeable bases (cmol/kg)	2-3	5-7	8-12
Exchangeable hydrogenand / or alluminium (cmol/kg)	4-8	1-2	0
Cation exchange capacity (cmol/kg)	12-18	20-26	20-26
Base saturation (%)	66.6	90-95	100
Probable pH	5.6-5.8	47	8-10

Source: Brady and Weil. 2002. Elements of the Nature and Properties of Soil. Prentice Hall, New Jersey.

Anion exchange

In the tropics, many highly weathered soils can have an anion exchange capacity. This means that the soil will attract and retain anions, rather than cations. In contrast to cations, anions are negatively charged. The anions held and retained by soil particles include phosphate, sulfate, nitrate and chlorine (in order of decreasing strength). In comparison to soils with cation exchange capacity, soils with an anion capacity have net positive charge. Soils that have an anion exchange capacity typically contain weathered kaolin minerals, iron and aluminum oxides, and amorphous materials. Anion exchange capacity is dependent upon the pH of the soil and increases as the pH of the soil decreases.

Base Saturation

Base saturation is a measurement that indicates the relative amounts of base cations in the soil. By definition, it is the percentage of calcium, magnesium, potassium and sodium cations that make up the total cation exchange capacity. For example, a base saturation of 25% means that 25% of the cation exchange capacity is occupied by the base cations. If the soil does not exhibit an anion exchange capacity, the remainder 75% of the CEC will be occupied by acid cations, such as hydrogen and aluminum. Generally, the base saturation is relatively high in moderately weathered soils that formed from basic igneous rocks, such as the basalts. The pH of soil increases as base saturation increases. In contrast, highly weathered and/or acidic soils tend to have low base saturation.

Movement of nutrient from soil to root

There are three basic methods in which nutrients make contact with the root surface for plant uptake. They are root interception, mass flow, and diffusion (Table 43).

- **Root interception**: Root interception occurs when a nutrient comes into physical contact with the root surface. As a general rule, the occurrence of root interception increases as the root surface area and mass increases, thus enabling the plant to explore a greater amount of soil. Root interception may be enhanced by mycorrhizal fungi, which colonize roots and increases root exploration into the soil. Root interception is responsible for an appreciable amount of **calcium** uptake, and some amounts of **magnesium, zinc** and **manganese**.
- **Mass flow:** Mass flow occurs when nutrients are transported to the surface of roots by the movement of water in the soil (i.e. percolation, transpiration, or evaporation). The rate of water flow governs the amount of nutrients that are transported to the root surface. Therefore, mass flow decreases are soil water decreases. Most of the **nitrogen, calcium, magnesium, sulfur, copper, boron, manganese** and **molybdenum** move to the root by mass flow.
- **Diffusion:** Diffusion is the movement of a particular nutrient along a concentration gradient. When there is a difference in concentration of a particular nutrient within the soil solution, the nutrient will move from an area of higher concentration to an area of lower concentration. You may have observed the phenomenon of diffusion when adding sugar to water. As the sugar dissolves, it moves through parts of the water with lower sugar concentration until it is evenly distributed, or uniformly concentrated. Diffusion delivers appreciable amounts of **phosphorus, potassium, zinc**, and **iron** to the root surface. Diffusion is a relatively slow process compared to the mass flow of nutrients with water movement toward the root.

Table 43 : Relative significance of the principal ways in which plant nutrient ion move from the soil to the roots of Corn

Nutrient	Amount of nutrient required for 150 bula of corn (lb/a)	Percent supply by Root interception	Mass flow	Diffusion
Nitrogen	170	1	99	0
Phosphorus	35	3	6	94
Potassium	175	2	20	78
Calcium	35	171	429	0
Magnesium	40	38	250	0
Sulphur	20	5	95	0
Copper	0.1	10	400	0
Zinc	0.3	33	33	33
Boron	0.2	10	350	0
Iron	1.9	11	53	37
Manganese	0.3	33	133	0
Molubdenum	0.01	10	200	0

Source: Barber, Stanley A. Soil bionutrient availability. John Wiley & Sons, New York, NY (1984)

Nutrient uptake into the root and plant cells

Before both water and nutrients are incorporated into plants, both must first be absorbed by plant roots.

Uptake of water and nutrients by roots

- Root hairs, along with the rest of the root surface, are the major sites of water and nutrient uptake.
- Water moves into the root through osmosis and capillary action.
- Soil water contains dissolved particles, such as plant nutrients. These dissolved particles within soil water are referred to as solute. Osmosis is the movement of soil water from areas of low solute concentration to areas of high solute concentration. Osmosis is essentially the diffusion of soil water.
- Capillary action results from water's adhesive (attraction to solid surfaces) and cohesion (attraction to other water molecules). Capillary action enables water to move upwards, against the force of gravity, into the plant water from the surrounding soil.
- Nutrient ions move into the plant root by diffusion and cation exchange.
- Diffusion is the movement of ions along a high to low concentration gradient.
- Cation ion exchange occurs when nutrient cations are attracted to charged surface of cells within the root, called cortex cells. When cation exchange occurs, the plant root releases a hydrogen ion. Thus, cation exchange in the root causes the pH of the immediately surrounding soil to decrease.

- Once water and nutrient ions enter the plant root, they move though spaces that exist within the root tissue between neighboring cells.
- Water and nutrients are then transported into the xylem, which conducts water and nutrients to all parts of the plant.

Once water and nutrients enter the xylem, both can be transported to other parts in the plant where the water and nutrients are needed. The basic outline of how nutrient ions are absorbed by plant cells follows.

Absorption of nutrients into plant cells

- Plant cells contain barriers (plasma membrane and tonoplast) that selectively regulate the movement of water and nutrients into and out of the cell. These cell barriers are permeable to oxygen, carbon dioxide, as well as certain compounds and semi-permeable to water. Wheras, selectively permeable to inorganic ions and organic compounds, such as amino acids and sugars.
- Nutrient ions may move across these barriers actively or passively. Passive transport is the diffusion of an ion along a concentration gradient. When the interior of the cell has a lower concentration of a specific nutrient than the outside of the cell, the nutrient can diffuse into the cell. This type of transport requires no energy. Wheras, active transport is the movement of a nutrient ion into the cell that occurs *against* a concentration gradient. Unlike passive transport, this type of movement requires energy.

Mobility of nutrients in soil

The knowledge of the mobility of micronutrients in plants helps in finding the deficiency of particular nutrient. Mobility of nutrient in soil is given in table: 1

Mobile:	The mobile micronutrients are highly soluble and are not adsorbed on clay complex.
Less mobile	They are also soluble but adsorbed on clay complex and so their mobility is reduced.
Immobile	These ions are highly reactive and get fixed in the soil.

Nutrient mobility in plant

The location of symptoms of nutrient deficiencies within plants will depend on the extent and rate of retranslocation of the nutrient from old leaves to new growth. Nutrients differ markedly in their mobility within the plant.

Mobile	Retranslocated from old leaves to new growth under all conditions

Variable mobility	Retranslocated from old leaves to new growth only under some conditions
Immobile	Not retranslocated from old leaves to new growth under any conditions

Nutrients such as nitrogen, phosphorus and potassium are readily translocated from old leaves to new growth. Hence, symptoms of these deficiencies occur initially in the older leaves. Nutrients such as calcium and boron do not appear to be retranslocated from old leaves to new growth under any circumstances. Hence, for these nutrients, deficiency symptoms occur generally in young growing areas of the plant.

For many nutrients, the extent of retranslocation is variable, depending upon the degree of the deficiency, the plant species, and the nitrogen status of the plant. For example, there is little or no movement of copper, zinc, or molybdenum out of old leaves of deficient plants. For these nutrients, symptoms will occur mainly in young tissues.

For nutrients where nitrogen supply affects the movement of other nutrients from old leaves to new growth, the location of symptoms may vary with fluctuation in the nitrogen supply. An example of this is the effect of nitrogen supply on the location of sulfur deficiency. In plants with an abundant supply of nitrogen, symptoms of sulfur deficiency occur initially on young leaves. However, in plants with a marginal supply of nitrogen, sulfur-deficient plants are generally pale and symptoms occur initially in the older leaves.

9

Methods of Fertilizer Application

In order to get maximum benefit from manures and fertilizers, they should not only be applied in proper time and in right manner but any other aspects should also be given careful consideration. Different soils react differently with fertilizer application. Similarly, the N, P, K requirements of different crops are different and even for a single a crop the nutrient requirements are not the same at different stages of growth. The aspects that require consideration in fertilizer application are listed below:

- Availability of nutrients in manures and fertilizers.
- Nutrient requirements of crops at different stages of crop growth.
- Time of application.
- Methods of application, placement of fertilizers.
- Foliar application.
- Crop response to fertilizers application and interaction of N, P, and K.
- Residual effect of manures and fertilizers.
- Crop response to different nutrient carrier.
- Unit cost of nutrients and economics of manuring.

Fertilizers are applied by different methods mainly for 3 purposes:

- Make the nutrients easily available to crops,
- Reduce fertilizer losses and
- Ease of application.

The time and method of fertilizer application vary in relation to

- The nature of fertilizer.
- Soil type and
- The differences in nutrient requirement and nature of field crops.

Application of fertilizers in solid form

I) Broadcasting

Even and uniform spreading of manure or fertilizers by hand over the entire surface of field while cultivation or after the seed is sown in standing crop, termed as broad casting. Depending upon the time of fertilizer application, there are two types of broadcasting:

A) **Broadcasting at planting:** Broadcasting of manure and fertilizers is done at planting or sowing of the crops with the following objectives:

- To distribute the fertilizer evenly and to incorporate it with part of, or throughout the plough layer and
- To apply larger quantities that can be safely applied at the time of planting/ sowing with a seed-cum-fertilizer driller.

It is adopted with the following condition:

- When nitrogenous fertilizers like ammonium sulphate, ammonium sulphate nitrate, concentrated organic manures, are to be applied to the soil deficient in N or where N is exhausted by previous crops like fodder, *Jowar*, F. maize.
- When citrate soluble phosphotic fertilizers like basic slag and dia-calcium phosphate, are to be applied to moderately acid to strongly acid soils.
- When potassic fertilizers like muriate of potash and potassium sulphate are to be applied in potash deficient soil.

B) **Top dressing:** Spreading or broadcasting of fertilizers in the standing crop after emergence of crop is known as top-dressing. Generally, NO_3^- N fertilizers are top dressed to the closely spaced crops like wheat, paddy e.g. sodium nitrate, ammonium nitrate and urea, so as to supply N in readily available from the growing plants. The term side dressing refers to the fertilizer placed beside the rows of a crop which grown widely spaced like maize or cotton. Care must be taken in top dressing that the fertilizer is not applied when the leaves are wet or it may burn or scorch the leaves. The top dressing of P and K is ordinarily done only on pasture lands which occupy the land for several years.

In some countries, aero planes are used for fertilizer application in hill terrains where it is difficult to transport fertilizers and where large amount are to be applied because of severe deficiency and under following situations:

- Where very small quantities of fertilizers are needed over large areas. e.g.: micro nutrients.
- When high analysis materials are applied.
- When fertilizer application may be combined with insect control or some other air operation and
- As a labour and time saving device.

II) Placement

In this, the fertilizers are placed in the soil irrespective of the position of seed, seedling or growing plant before or after sowing of the crops. It includes:

1. **Plough sole placement:** The fertilizer is placed in a continuous band on the bottom of the furrow during the process of ploughing. Each band is covered as the next furrow is turned. By this method, fertilizer is placed in moist soil where it can become more available to growing plants during dry seasons. It results in less fixation of P and K than that which occurs normally when fertilizers are broadcast over the entire soil surface.
2. **Deep placement or sub-surface placement:** In this method, fertilizers like Amm. Sulphate and Urea, is placed in the reduction zone as in paddy fields, where it remains in ammonia form and is available to the crop during the active vegetative period. It ensures better distribution in the root zone, and prevents any loss by surface runoff. It is followed in different ways, depending upon local cultivation practices such as:

i) **Irrigated tracts:** The fertilizer is applied under the plough furrow in the dry soil before flooding the land and making it ready for transplanting.

ii) **Less water condition:** Fertilizer is broadcasted before puddling which places it deep into the reduction zone.

iii) **Sub-soil placement:** This refers to the placement of fertilizers in the sub-soil with the help of heavy power machinery. It is followed in humid and sub-humid regions where many sub-soils are strongly acid, due to which the level of available plant nutrients is extremely low. Phosphotic and potassic fertilizers are applied by this method in these regions for better root development.

III) Localized placement

It refers to the application of fertilizers into the soil close to the seed or plant. It is usually employed when relatively small quantities of fertilizers are to be applied. It includes methods like:

Advantages

i) The roots of the young plant are assured of an adequate supply of nutrients,

ii) Promotes a rapid early growth,

iii) Make early intercultivation possible for better weed control,

iv) Reduces fixation of P and K.

1. **Contact placement or combined drilling or drill placement:** It refers to the drilling of seed and fertilizer together while sowing. It places the seed and small quantities of fertilizers in the same row. This is found useful in cereal crops, cotton and grasses but not for pulses and legumes. This may affect the germination of the seed, particularly in legumes due to excessive concentration of soluble salts.
2. **Band placement:** In this, fertilizer is placed in bands which may be continuous or discontinuous to the side of seedling, some distances away from it and either at level with the seed, above the seed level or below the seed level. There are two types of band placement: It includes hill and row placement.

a. **Hill placement:** When the plants are spaced 3 ft. or more on both sides, fertilizers are placed close to the plant in bands one or both sides of the plants. The length and depth of the band and its distance from plant varies with the crop and the amount of fertilizer as in cotton.

b. **Row placement:** When the seeds or plants are sown close together in a row, the fertilizer is put in continuous band on one or both sides of the one or both sides of the row by hand or a seed drill. It is practiced for sugarcane, potato, maize, tobacco, cereals and vegetable crops.

Higher rates of fertilizers are possible with row placement than hill placement. For applying small amount of fertilizers, hill placement is usually most effective.

3. **Pellet application:** In this method, fertilizer (N-ous fertilizers) is applied in the form of pellets 2.5 – 5.0 cm. deep between the rows of paddy crop. Fertilizer is mixed with soil in the ratio of 1:10 and make small pellets of a convenient size are then made and deposited in the soft mud of paddy fields. It increases the efficiency of N-ous fertilizers.
4. **Side dressing:** Fertilizers are spread in between the rows or around the plants. It includes i) application of N-ous fertilizers in between the rows by hand to broad row crops like maize, S.cane tobacco, cereals which is done to supply additional doses of N to the growing crop. ii) Application of mixed or straight fertilizer around the base of the fruit trees and done once, twice or thrice in a year depending upon age.

10

Application of Biofertilizers in Vegetables

Biofertilizers are preparations containing micro-organisms, with capability of mobilizing nutritive elements from non usable form to usable form through biological processes. Bio-fertilizers have definite advantage over chemical fertilizers. Chemical fertilizers supply over nitrogen whereas bio-fertilizers provide in addition to nitrogen, certain growth promoting substances like hormones, vitamins, amino acids, *etc*, crops have to be provided with chemical fertilizers repeatedly to replenish the loss of nitrogen utilized for crop growth. Bio-fertilizers application in agriculture will have greater impact on organic agriculture and also on the control of environmental pollution, soil health improvement and reduction in input use. Farmers try to increase yields of vegetable crops by mean of heavy nutrition's. The use of chemical nitrogen and phosphorus fertilizers at high levels had an adverse effect on the accumulation of NH_4+, NO_3-, NO_2- and PO_4- in vegetable product tissues. Therefore, clean agriculture recently depends upon using biofertilizers as well as organic in order to produce high yields with the best commodity quality without contamination and less accumulation with heavy metals.

The application of high input technologies such as chemical fertilizers, pesticides, herbicides have improved the production but there is growing concern over the adverse effects of the use of chemicals on soil productivity and environment quality. Thus, integrated nutrient management has become an accepted strategy to bring about improvement in soil fertility and protecting the environment. This strategy utilizes a judicious combination of fertilizers, organic manures and bio-fertilizers. Biofertilizer is a natural product carrying living microorganisms derived from the root or cultivated soil. So they don't have any ill effect on soil health and environment. Besides their role in atmospheric nitrogen fixation and phosphorous solubilisation, these also help in stimulating the plant growth hormones providing better nutrient uptake and increased tolerance towards drought and moisture stress. A small dose of biofertilizer is sufficient to produce desirable results because each gram of carrier of bio-fertilizers contains at least 10 million viable cells of a specific strain. Biofertilizers means the product containing carrier base (solid or liquid) living microbial preparations which are agriculturally useful in terms of nitrogen fixation, phosphorus solubilization or nutrient mobilization to increase the productivity of the soil and/or crop. These preparations in strict terms are called as microbial inoculants. Biofertilizers a cost effective renewable energy source play a crucial role in reducing the inorganic fertilizer level and at the same time increasing the crop yield besides maintaining the soil fertility. The replenishment of nutrients lost by crop removal through the use of chemical fertilizers alone is not advisable in the long run, since their continuous use, impaired the soil health and productivity. Biofertilizers can help in reducing the input of inorganic fertilizers for an extent of 25 per cent for obtaining the same or higher yield. Biofertilizers are microbial preparations containing live or latent cells of specific microorganisms to apply to the seed or soil, so that the cells can multiply and bringing out the activities of nitrogen fixation or phosphate solubilization / mobilization in the root region or in association with the root system of crop plants.

On the other hand biofertilizers supply the nitrogen continuously throughout the entire period of crop growth in the field under favourable conditions. Continuous use of chemical fertilizers adversely affects the soil structure whereas biofertilizers when applied to soil improve the soil structure.

The effects of chemical fertilizers are that they are toxic at higher doses. Biofertilizers, however, have no toxic effects. Biofertilizers are commonly called as microbial inoculants which are capable of mobilizing important nutritional elements in the soil from non-usable to usable form by the crop plants through their biological processes. Chemical fertilizers are expensive; they disturb the equilibrium of agro-ecosystems and cause pollution to the environment. The utilization of microbial products has several advantages over conventional chemicals for agricultural purposes:

- Microbial products are considered safer than many of the chemicals now in use.
- Neither toxic substances nor microbes themselves will be accumulated in the food chain.
- Self-replication of microbes circumvents the need for repeated application.
- Target organisms seldom develop resistance as is the case when chemical agents are used to eliminate the pests harmful to plant growth.
- Properly developed bio control agents are not considered harmful to ecological processes or the environment.

The deleterious effects of chemical fertilizers are that they are toxic at higher doses. Many of these countries are attempting to farm marginal land that requires an input of nutrients. Current costs for chemical fertilizer are so high that farmers in many developing cannot afford them, and must find alternative nutrient inputs. But biofertilizers are no substitute for chemical fertilizers.

Biofertilizers are known to make a number of contributions in agriculture

- They supplement fertilizer supplies for meeting the nutrient needs of crops.
- They can add 20-200 kg N/ha (by fixation) under optimum conditions and solubilize/mobilize 30-50 kg P_2O_5/ha.
- They liberate growth promoting substances and vitamins and help maintain soil fertility.
- They suppress the incidence of pathogens and control diseases. They increase crop yield by 10-15%, N-fixers reduce depletion of soil nutrients and provide sustainability to the farming system.
- They improve soil physical properties, tilth and soil health in general. Microorganisms such as bacteria, fungi and actenomyces play a principle role in N fixation and P availability in soil which may increase the uptake of N and P through plant roots. These microorganisms could be grown under laboratory conditions and then applied to seeds, roots or directly to the soil. The aim of using N-biofertilizers is to increase soil content of free living bacteria such as; *Azotobacter* sp., *Azospirillum* sp., *Klebsiella* sp. and others which are expected to increase N- fixation in the soil. Of course symbiotic bacteria of genus *Rhizobium* is also considered as a good way of N- fixation in legume crops. Among the various nutrients, nitrogen is one of the key elements for higher productivity. Nitrogenous fertilizers and biological nitrogen fixation represent the major inputs of nitrogen for crops. Although improved technologies of fertilizer nitrogen production and increased efficiency of fertilizer use by plants could make more nitrogen available for the plants, alternate technologies are being sought to reduce the dependence of plants on fertilizer nitrogen. The nitrogen-fixing microorganisms like *Rhizobium, Azospirillum, Azotobacter, Azolla,* cyanobacteria etc. fixes nearly 175 x 106 tonnes of N on the earth surface and it may be possible to meet a large part of nitrogen demand through proper manipulation of these organisms in crop production system. The use of inoculants as alternatives to N fertilizer avoids problems of contamination of water resources from leaching and runoff of excess fertilizer.

***Efficiency*:** Legume inoculants do not require high levels of energy for their production or distribution. Application on the seed is simple compared to spreading fertilizer on the field.

***Better yields*:** Inoculants increase legume crop yields in many areas. BNF often improves the quality of dietary protein of legume seed even when yield increases are not detected.

***Increased soil fertility*:** Through practices such as green manuring, crop rotations, and alley cropping, N fixing legumes can increase soil fertility, permeability, and organic matter to benefit non-legume crops.

***Sustainability*:** Using BNF is part of the wise management of agricultural systems. The economic, environmental, and agronomic advantages of BNF make it a cornerstone of sustainable agricultural systems.

Azotobacter: *Azotobacter* represents the main group of heterotrophic free living nitrogen-fixing bacteria. They are Gram negative, large ovoid pleomorphic cells of 1.5-2.0 ìm or more in diameter ranging from rods to coccoid cells. They occur singly, in paired or irregular clumps and sometime in chains of varying length. They do not produce endospores but form cysts. They are motile by peritrichous flagella or non motile.

Azotobacter spp. is most specifically noted for their nitrogen fixing ability but they have also been noted for their ability to produce different growth hormones (IAA and other auxins, such as gibberllins and cytokinins), vitamins and siderophores. *Azotobacter* is capable of converting nitrogen to ammonia, which in turn is taken up by the plants. *Azotobacter* sp. can also produce antifungal compounds to fight against many plant pathogens.

Table 44 : Response of vegetable crops to bio-fertilizer inoculations (for nitrogen)

Bio-fertilizer	Crop	Increase in yield (%)	Nitrogen economy (%)	References
Rhizobium	Pea	13.38	-	Kanaujia *et al.* (1999)
	Pea	5.10	-	Choudhury *et al.* (1982)
Azotobacter	Cabbage	24.30	25	Verma *et al.* (1997)
	Garlic	14.23	25	Anonymous (2003)
	Knol khol	9.60	25	Chatto *et al.* (1997)
Azospirillum	Cabbage	11.87	25	Verma *et al.* (1997)
	Onion	21.68	25	Anonymous (2002)
	Radish	9.00	-	Sundaravelu and Mutukrishna (1993)
	Sweet potato	8.50	-	Desmond *et al.* (1990)

Table 45: Yield increase in vegetables at farmers fields at different locations

Place	Treatment	Crops	Yield		Increase in yield over untreated (%)
			Control (q/ha)	Treated (q/ha)	
Umri Nagpur	Azotobacter	Okra	24.8	26.0	8.3
Ambada Narkhed	PSM	Brinjal	125.0	137.5	10.0
Tivara Amravati	Azotobacter	Brinjal	190.0	220.0	15.8
Nagapur Sweagram	Azotobacter	Chilli	14.5	16.0	10.3
Bopapur Nagpur	PSM	Cauliflower	34.0	36.5	7.35
Chikhali Katol	Azotobacter	Cauliflower	32.5	34.5	6.2
Chicholi Parshivani	Azotobacter	Okra	23.4	25.5	8.97

Source: RBDC, VCA Complex, Nagpur (1997-98).

Table 2 : Response of vegetable crops to bio-fertilizer inoculations (for phosphorus)

PSM	Garlic	14.23	25	Anonymous (2003)
	Pumpkin	51.00	25	Karuthamani *et al.* (1995)
VAM	Onion	4.70	25	Gurubatham *et al.* (1989)

Azospirillum

They are called as associative endosymbiont on roots of grasses and similar types of plants. They are also known to fix atmospheric nitrogen and benefit host plants by supplying growth hormones and vitamins. Generally the nitrogen requirement of non-leguminous crops such as cereals, grasses and horticultural crops met partly from the activities of associative symbiotic bacteria-*Azospirillum*. *Azospirillum* is considered to be more efficient and it has been reported that *Azospirillum* inoculation increases the growth, nitrogen uptake and yield in number of crops. Number of scientific findings were well documented that application of *Azospirillum* and other plant growth promoting rhizobacteria improve the plant growth and yield of commercially important crops like tomato, brinjal and chillies (Sukhada 1988). Kumaran *et al.* (1998) reported that tomato growth, yield and quality parameters such as TSS, ascorbic acid and lycopene contents were higher in plants grown with phosphobacteria and *Azospirillum*.

Phosphobacteria : Microorganisms are also involved in the availability of phosphorus, the second most important nutrient required by crop plants. The phosphate solubilizing bacteria

(PSB) solubilize the insoluble phosphates and make them available for crop plants in the rhizosphere region. Several soil bacteria and fungi notably species of *Pseudomonas,Bacillus, Penicillium* and *Aspergillus* etc. secrete organicacids and lower the *p*H in their vicinity to bring about solubilization of bound phosphates in soil. Increased in the yield of various crops were demonstrated due to inoculation of peat based cultures of phosphobacteria and saving up to 50 per cent of recommended level of P_2O_5 was observed in many experiments. The chilli cultivars (Arka lohit, Arka jwala and Arka anamika) and Brinjal inoculated with Phosphobacteria found to increase crop growth and yield (Sukhada, 1999).

Vesicular Arbuscular Mycorrhiza (VAM) Mycorrhiza is the mutualistic association between plant roots and fungal mycelia. Many graminaceous plants legumes and horticultural crops are highly susceptible to VAM colonization. The transfer of nutrients mainly phosphorus from the soil to the cells of the root cortex is mediated by intracellular obligate fungal endosymbiont of the genera *Glomus, Gigaspora, Acaulospora, Scleroscystis:* and *Endogone* which posses vesicles for storage of nutrients and arbuscules for funneling these nutrients into the root system. The mycorrhizal fungi mobilize phosphates and other micro nutrients like zinc, boron and molybdenum from adjacent soil to the root system through hyphal network.

Enhanced uptake of phosphorus and increased plant growth due to inoculation of soil with VAM fungi in horticultural crops such chilli (Bagyaraj and Sreeramulu 1982), tomato (Fairweather and Parbery, 1982), asparagus (Hussey *et al.,* 1984), potato (Bhattarai and Mishra, 1984) and lettuce (Water and Coltman, 1989) has been observed.

Combined inoculation

The highest number of fruits, fruit weight, length of fruits and thickness of fruits were obtained with the application of combination of organic manures together with *Azotobacter* and PSB in okra crop (Bairwa *et al.*, 2009). Gomaa (1995) mentioned that seeds inoculation of tomato plants with a mixture of *Azotobacter chroococcum, Azospirillum brasilense* and *Bacillus subtilus* results an increase in fresh and dry weight of plants over inoculating plants with *Azospirillum brasilense* or *Azotobacter chroococcum* alone. Murumkar and Patil (1996) studied the effect of *vesicular arbuscular mycorrhizal* (VAM) fungi and its combination with diazotrophs *Azotobacter chroococcum, Azospirillum lipoferum* on *Capsicum annuum* cv. California Wonder. They found that plant dry weight significantly increased by mycorrhizal inoculation together with different diazotrophs. Khan and Pariari (2012) showed that the inoculation with Azospirillum + 75% N along with P and K recorded maximum fruit yield (10.25 t/ha), number of fruits/plant (80.20), fruit length (6.72 cm) and oleoresin content (19.80%). All the growth characters like plant height (62.75 cm) and number of branches / plant (25.15) was found highest with Azospirillum + 100% N + PK, which were *at par* with Azotobacter + 100% N + PK. Allthe treatments were found superior over control (recommended dose of NPK only). Considering the cost: benefit ratio and yield of the crop, Azospirillum was considered better over Azotobacter.

Table 46 : Typical levels of nitrogen fixation from different systems

Crop/Plant	Associated crops	Typical levels of nitrogen fixation (kgN/ha) per year
Symbiotic		
Legume (nodulated)	**Bacteria (Rhizobium)**	
Alfalfa (*Medicago sativa*)		150-250
Clover (*Trifolium pretense L*)		100-150
Soyabean (*Glysine max* L.)		50-150
Cowpea (*Vigna unguiculata*)		50-100
Lupine (*Lupines*)		50-100
Vetch (*Vicia vilbosa*)		50-125
Bean (*Phaseolus vulgaris*)		30-50
Nonlegume *(nodulated)*		
Alders (Alnus)	Actinomycetes (Frankia)	50-150
Species of Gunnera	Blue green algae (Nostok)	10-20
Nonlegume *(Non-nodulated)*		
Pangola grass (Degetaria decumbens)	Bacteria (Azosprilium)	5-30
Bahia grass (Paspalum notatum)	Bacteria (Azotobactor)	5-30
Azolla	Blue-green algae (Anabena)	150-300
Nonsymbiotic	Bacteria (Azotobactor, clostridium)	5-20
	Blue-green algae (various)	10-50

Source: Brady and Weil. 2002. Elements of the Nature and Properties of Soil. Prentice Hall, New Jersey.

How to apply biofertilizers in vegetables

There are methods for applying bio-fertilizers in vegetables i.e seed treatment, cut piece/set treatment, seedling treatment and soil application.

Seed Treatment

- About 200 g of bio-fertilizers is required to treat 10-14 kg of seed. Suspend one packet of 200g in approximately 400 ml water and mix it thoroughly.
- Pour this mixture on seeds and mix with hands to obtain uniform coating on each and every seed than spread the seeds in shade for drying for 10-15 minutes then sow them immediately.

Cutting/Set treatment

- Prepare a culture suspension by mixing 1 kg of culture in 50-60 litres water and the cut pieces of planting material required for 1 acre are kept immersed in the suspension for 10-15 minutes.
- Then bring out these cut pieces and allow to dry for sometimes before planting. This method is applicable for crops like potato.

Seedling treatment

- Seedling treatment is recommended for tomato, chilli, onion etc.
- Prepare the suspension by mixing 1 kg of culture in 10-15 litres of water and get seedlings required for 1 acre and make small bundles of seedlings.
- Dip the seedlings in the suspension for 15-20 minutes and than transplant these immediately.
- Generally the ratio of inoculants and water should be 1:10 approximately, i.e. 1 kg packet in 10 litres of water.

Soil Application

- Prepare the mixture of 2-3 kg of bio-fertilizer in 40-60 kg of soil/compost. Broadcast the mixture in one acre of land either at sowing time or 24 hr before sowing. The application of phosphate-solubilizers is very common.

Effect of locally available organic manures, green manures and biofertilizers in different combinations improve growth and yield of vegetables which can improve the economics of cultivation of farmers. The potential nitrogen fixing biological systems like *Azospirillum*, phosphorus-solubilizing bacteria such as *Bacillus* and *Pseudomonas* and the vesicular arbuscular mycorrhizae plays a triggering role in nitrogen and phosphorus nutrition of vegetable crops under protected cultivation. Application of biological fertilizer greatly involved in the accumulation of soil enzymes, which directly reflects on soil fertility index. The effective utilization of bio-fertilizers for crops not only provide economic benefits to the farmers but also improves and maintain the soil fertility and sustainability in natural soil ecosystem.

11

Factors Affecting of Fertilizer Application or Requirement

Factors for deterring the Fertilizer Schedule are two factor

Soil Supplying Power: Growing different crops during different seasons alters the soil nutrient status, estimated by soil analysis at the beginning of the season. The soil supplying power increases with legume in rotation, fertilizer application and addition of crop residues. The available nitrogen and potassium in soil after groundnut are higher to initial status of the soil. But after pearl millet, only potassium status in the soil is improved and no changes in P.

Nutrient Uptake by Crops: The total amount of nutrients taken by the crops in one sequence gives an indication of the fertilizer requirement of the system. The balance is obtained by subtracting the fertilizer applied to crops that nutrient taken by the crops.

Residual Effect of Fertilizers: The extent of residues left over in the soil depends on the type of fertilizer used. Phosphatic fertilizer and FYM have considerable residue in the soil, which is useful for subsequent crops. The residues left by potassium fertilizers are marginal. Legumes add nitrogen to the soil in the range of 15 to 20 kg/ha. The amount of nitrogen added depends on the purpose. Green gram grown for grain, contributes 24 and 30 kg N respectively to the succeeding crop. Inclusion of leguminous green manures in the system add 40 kg to 120 kg N/ha. The availability of phosphorous is also increased by incorporation of green manure crops. Potassium availability to subsequent crop is also increased by groundnut crop residues: crop residues add considerable quantity of nutrients to the soil, cotton planted in finger millet stubbles benefits by 20 to 30 kg/ha due to decomposition of stubbles. Deep rooted crops- cotton, red gram absorbs nutrients from deeper layers. Leaf fall and decomposing add phosphorus to top layers crop residue contain high C: N ratio like stubbles of sorghum, pearl millet temporarily immobile nitrogen. Residue of legume's crop contains low C: N ratio and they decompose quickly and release nutrients.

Efficiency of crops: Jute is more efficient crop for utilizing of nitrogen followed by summer rice, maize, potato and groundnut in that order. Phosphorus efficient crops, jute > summer rice> *Kharif* rice> potato > groundnut > maize. Groundnut is more efficient in potassium utilization followed by maize, jute, summer rice, *Kharif* rice, and potato.

Fertilizer recommendation should be based on cropping system e.g. in wheat based cropping system an extra dose of 25% nitrogen is recommended for wheat when it is grown after sorghum, pearl millet. When wheat, after pulse crop needs 20 to 30 kg less nitrogen per hectare. Phosphatic fertilizers are added through green manure crops, not to apply phosphates to succeeding wheat crop. In rice based cropping system consisting of rice- rice in *Kharif* and *rabbi* and sorghum, maize, finger millet, soybean in summer it is sufficient to apply phosphorus and potassium to summer crops only while nitrogen is applied to all the crops. Thus, following system approach in fertilizer recommendation can save lot of fertilizer.

Water management: There is no carry over effect of irrigation as in case of fertilizer, rice – rice is efficient cropping system for total yield, but it consume large amount of water especially in summer. If water is scare in summer instead of rice, groundnut is used in cropping system.

Method of irrigation: The layout should be so planned that most of the crops can be suitable, in rice- rice- groundnut system; rice is irrigated by flood method, while groundnut by boarder strips. In cotton – sorghum- finger millet system, cotton, sorghum by furrow method while finger millet checks – basin method is adopted.

More remunerative and less water consuming crop rotation have standardized have been standardized at different location of India. Rice- mustard-green gram, rice- potato- green gram rotation were found more water efficient systems at memari in Memari in W.B., under high level of irrigation in tarai region of U.P, rice-lentil and rice-wheat cropping system were found better. Pre monsoon groundnut-rabi sorghum sequence was highly remunerative with high water use efficiency compared to sugarcane alone in M.S when irrigation water is not limiting. Under limited water supply, however, rice – chickpea- green gram and rice- mustard – green gram are more remunerative with high water use efficiency.

Weed management: Weed problems observed individual crop, and weed shift occur and their carry over effect of weed control method on succeeding crop.

Weeds are dynamic in nature, generally broad- leaved weeds occur in wheat occur in wheat at later stage and 2, 4 D is applied as post emergence herbicide to control them. In rice-wheat system, canary grass (phallaris minor) is a menace for wheat crop. Other weed seed species are decomposed and loss viability, but Phataris minor seed do not loss viability. When sown in rice stubble is heavily infested with Phalaris minor.

Herbicide applied to the previous crop may be toxic to the succeeding crop. Higher dose of Atrazine applied to sorghum crop affect germination of succeeding pulse crops. Herbicide recommendation should be depends on succeeding crops, ploughing before the planting season helps to kill most of the weeds.

Pest and Diseases: Pest and diseases infestation more in sequence cropping due to continuous cropping, carry over effect of insecticides is not observed.

Harvesting: In sequences cropping crop can be harvested at physiological maturity stage instead of harvesting maturity. The field can vacate one week earlier. Because of continuous cropping the harvesting time may coincide with heavy rains and special post harvest operations, like artificial drying, treating the crop with common salt etc. are practices to save the produce.

Integrated farming system, components and its advantages.

India with 2.2 percent of global geographical area support more than 15 percent of the total world population, 70 percent of who depends on agriculture. It is also support nearly 15 percent of the total livestock population of the world.

As of now, out of 328.73 ha of geographical areas approximately 18 percent is forest only 13.5 percent is not available for cultivation. Total problems areas constitute 173.65 million ha, which include areas subjected to wind and water erosion (145 million ha) water-logged areas (8.53 million ha), alkali soils (3.58 million ha), saline and coastal sandy areas (5.50 million ha), ravines and gullies (3.97 million ha), shifting cultivation (4.91 million ha) and reverie torrents 2.73 (million ha). Besides 40 million ha are prone to flood and 260 million ha. Is drought prone. Thus the net sown area is 136.18 million ha. Is drought prone. Thus, the net sworne is 136.18 million ha (41.42 percent of the total geographical area) (subbing *et al.*, 2000).

Unlike industries, agriculture is practiced by 105 million farm families who live in 0.6 million village. More than 40 percent of them are below the poverty line. Nearly 85 million farm families belong to small and marginal categories. Only 25 to 30 % of the modern agricultural technology has reached the farmers. This is often because the technology has not been consistent with the condition with condition of the farm situations. Since there is no further scope for horizontal expansion of land for cultivation, the only alternative left is for vertical expansion on space and time particularly for small and marginal farmers (constituting 76% farming community) who do not have much of reasons, especially in rain fed areas. The new farming system research strategy should, therefore to develop technology with participatory approach of farmers.

Nutrient deficiency indicator plants

Some of the crops are known to be specific for the occurrence of symptoms of a particular deficient nutrient element exhibiting characteristic symptoms. Such crops are called as indicator crops for the deficiency of that particular element(s). This is mainly due to the greater demand of the element in the respective Indicator Crops. Some of the Indicator Crops are furnished in the following Table 47 along with the elements.

Table 47 : Nutrient deficiency indicator plants

Nitrogen	Maize, Sorghum, Leguminous plants
Phosphorus	Tomato, Maize, Lucerne, Cereals, Duranta
Potassium	Maize, Lucerne, Cotton, Potatoes, Banana, Cucurbits
Sulpur	Lucerne, Clover, Cereals, Tea
Zinc	Maize, Tomatoes, Potatoes, Beans, Citrus
Copper	Citrus
Iron	Ornamental plants, Ixora, Acacia, Eucalyptus, Gooseberry, Securmanis
Boron	Lucerne, Coconut, Guava
Manganese	Citrus
Molybdenum	Cauliflower, Cabbage
Sodium	Sugar beet and Turnip

Source- http://agritech.tnau.ac.in/agriculture/agri_min_nutri_chart.html

12

Use of PSB/Phosphobacteria on Phosphorus Economy in Different Potato Growing Regions of India

Phosphorus is one of the key nutrients required for higher and sustained productivity of potato and its influence on tuber yield is very well established. Phosphatic fertilizers are expensive and in developing countries like India, they are either imported or manufactured using imported raw material. Due to increase in cost in the recent past, there has been a decreasing trend in the amounts of P fertilizer applied in India (Sundara and Natarajan, 1997 and Sundara, 1988). At the same time, soils contain substantial reserves of total P, most of it remains relatively inert, and only less than 10% of soil P enters the plant–animal cycle (Kucey *et al.*, 1989). Upon addition to the soils, soluble phosphates react with the constituents of the soil and form compounds that are less soluble. This conversion depends upon the soil type. In acid soils, the reaction products are aluminium and iron phosphates and in the predominantly calcareous soils, the reaction products are calcium phosphates. As a result most of the P applied (often as much as 90%) is rendered unavailable for crop uptake but is retained in insoluble form. This is the reason that the potato-based cropping systems in different potato growing pockets generally show positive P balance resulting into its build up (Singh *et al.*, 1994). Thus, soils commonly have large reserves of 'fixed' P that could support long-term crop requirements if it could be mobilized through appropriate soil management including use of P-solubilizing microbes. The ability of phosphate solubilizing bacteria (PSB) to convert insoluble forms of phosphorus to an accessible form is an important trait in sustainable farming for increasing plant yields (Singh and Kapoor, 1994). The beneficial effects of PSB on crop productivity have been widely described. These soil phosphate solubilizing bacteria stay near the roots and make the phosphorus available to plants from soil.

The extent of benefit from these micro-organisms depends on their population and efficiency which, in turn is governed by a soil and environmental factors.

Different potato growing agro-ecological zones are likely to very in the environmental factors resulting into differential responses to applied PSB. Keeping this in view, a multilocation experiment was conducted in the major potato growing states of India under the All India Coordinated Research Project on Potato to evaluate role of phosphorus solubilizing bacteria on P economy in potato. The exact quantification of phosphorus fertilizer in presence of PSB from mobilization soil P is important. Therefore, yield obtained at different levels of P in presence of PSB was fitted in quadratic response model.

Method of application

Seed tubers were inoculated with phosphobacteria culture purchased from the authentic source and dried in the shade before planting. Nitrogen was applied in two splits as par recommendation *i.e.* half at planting and rest at earthing up at 40 days after planting through calcium ammonium nitrate. Basal application of P and K was done using single super phosphate and muriate of potash respectively, as per treatments at the time of planting. The crop was harvested at maturity and tuber yield and numbers were recorded from all treatments. Data were analyzed separately for each center following standard statistical procedure. For working out net return, price of potato was taken as per respective region. For exact quantification of contribution of PSB in reducing the requirement of fertilizer P, tuber yield of different centers at varying levels of P in presence of PSB were fitted in quadratic response model ($Y = ax^2 + bx + c$) and taking yield obtained at recommended fertilizers was as target (T) and amount of P required (%) to achieved the same in presence of PSB was calculated using the relationship % P of recommended dose required in presence of PSB to obtain a target yield

$$(T) = \frac{-b \pm \sqrt{b^2 - 4a(c-T)}}{2a}$$

At Kalyani, in the first year, there was no significant difference in potato tuber yield due to different treatments (Table 48). However, maximum total tuber yield (27.09 t/ha) was observed with recommended NPK followed by recommended NK along with 75% of recommended P + phosphobacteria (26.71 t/ha). Whereas, maximum number of tubers (598.8 thousand/ha) were observed with recommended dose of NK + 25% of P and PSB (Tables 3). The recommended dose of NPK gave highest net returns (Rs. 46551/ha) followed by recommended dose of NK + 75% P + phosphobacteria (Rs. 44639/ha) and recommended dose of NK + 50% P + phosphobacteria (Rs. 44526/ha). During second year, the highest tuber yield of 27.19 t/ha was recorded in the treatment where crop was fertilized with recommended doses of NK along with 75% of P + seed tubers inoculation with phosphobacteria followed by recommended NPK (26.61 t/ha). The lowest yield was recorded in the control where only recommended dose of NK was applied. Similar trend was observed in net return with highest value in recommended NK + 75% P + phosphobacteria (Rs.33300/ha) followed by application of recommended NPK (Rs.29870/ha). Pooled data indicated maximum yield with recommended dose of NK + 75% P + phosphobacteria (26.95 t/ha) which was statistically at par with recommended dose of NPK (26.85 t/ha). Both the treatments were significantly better than control.

At Chhindwara, the recommended dose of NPK recorded significantly higher tuber yield (31.6 and 33.05 t/ha) during both the year which was at par with recommended dose of NK + 75% P + phosphobacteria (30.2 and 31.05 t/ha). Trend of yield in both year was also reflected in pooled yield and maximum tuber yield (32.34 t/ha) was observed in recommended dose of NPK followed by recommended dose of NK + 75% P + phosphobacteria (31.84 t/ha) which was significantly better than control. Recommended dose of NK (without recommended dose of P) produced significantly higher number of total tubers both the year as well as in the pooled analysis. The highest net return was observed in recommended dose of NPK (Rs. 82103 and 115571/ha) which was closely followed by recommended NK along with 75% P and PSB (Rs. 80784 and 115757/ha) and recommended NK + 50% P + PSB (Rs. 77870 and 109429/ha) during both the year.

At Kota, recommended dose of NPK during both the year recorded highest tuber yield and net returns (28.55 and 23.97 t/ha and Rs. 73161 and 54103/ha, respectively) which was followed by recommended dose of NK along with 75% P and phosphobacteria (27.43 and 22.0 t/ha and Rs. 69129 and 47139/ha). Pooled analysis also showed that maximum yield (26.27 t/ha) with recommended NPK followed by recommended NK + 75% P + phosphobacteria (24.72 t/ha).

At Patna, the highest total tuber yield was recorded with recommended NPK (36.02 and 30.87 t/ha) which was statistically at par with use of PSB and 75% P with recommended NK (35.89 and 30.66 t/ha) during both the year. Tuber production with recommended dose of NK + 75% P + phosphobacteria (33.26 t/ha) was statistically at par with recommended dose of NPK (33.44) in pooled analysis also. However, number of tubers was maximum with recommended dose of NK + 75% P + phosphobacteria (824.25 thousand/ha) and was at par with recommended NPK. Similar trend was observed in net returns from these two treatments (Rs. 63022/ha and Rs. 62673/ha and Rs. 26997/ha and Rs. 25513/ha, respectively during 2005-06 and 2006-07.

At Hassan, potato tuber yield as well as tuber numbers were also significantly higher in the treatment receiving recommended doses of N and K along with 75% of P and PSB (44.15 and 28.22 t/ha and 388.4 and 359.4 thousand/ha, respectively) followed by the treatment receiving recommended doses of N, P and K (37.71 and 216.98 t/ha and 255.2 and 340.3 thousand/ha). Pooled data also reflected trends observed for tuber yield and number during both the years of experiment. Maximum net return of Rs.1, 07,925 and 87,725/ha was also obtained with the same treatment followed by recommended doses of NPK (Rs. 99840 and 85000/ha). Similarly, at Ooty also, results reveal that phosphobacteria treatment could compensate 25% of P recommended dose as the yield and tuber numbers obtained with 100% recommended dose of P and 75% recommended P + phosphobacteria (21.78, 11.15 and 21.30, 9.82 t/ha) were at per during both year as well as in pooled result. Other treatments showed significantly less number of tubers. Net return (Rs.26,997/ha) was higher when recommended doses of NPK was applied to potato closely followed by the treatment in which 25% of phosphorus dose was substituted by phosphobacteria (Rs.25,513/ha) in the first year.

At Bhubaneshwar, there was no significant difference among various treatments in the first year of experiment but in second year, the highest total tuber yield and numbers of (22.5 t/ha and 500.2 thousand/ha) were recorded in recommended dose of NPK followed by use of phosphobacteria with 75% P along with recommended NK doses (20.58 t/ha) (Table 2 and 4).

Pooled data repeated the trend having yield of recommended dose of NPK and recommended dose of NK + 75% P + phosphobacteria at par and better than other treatments. Similar trend was observed for net returns for these two treatments (Rs. 29142/ha and 21850/ha for recommended dose of NPK and recommended dose of NK + 75% P + phosphobacteria, respectively). At Dholi, application of recommended dose of NPK produced highest yield (16.41 and 11.81 t/ha) as well as number of total tubers (402.4 and 394.9 thousand/ha) during both the year as well as pooled yield of two years. However, in tuber numbers during second year, recommended dose of NPK was significantly better than recommended dose of NK + 75% P + phosphobacteria, Similar trend was observe in net return, the highest net returns (Rs. 40950 and 29655/ha) was obtained for the recommended dose of NPK followed by recommended dose of NK along with 75% P and phosphobacteria (Rs. 36250 and 20270/ha) during both the years of experimentations.

At Hisar, during both the year, total tuber yield was highest in the treatment with recommended dose of NPK (30.55 and 31.65 t/ha) followed by recommended dose of NK + 75% P + phosphobacteria (29.51 and 28.95t/ha). Similar trend for tuber yield of both the year was also reflected in pooled analysis. Similarly, the highest net return was obtained with recommended dose of NPK (Rs. 65159 and 69635/ha) followed by recommended dose of NK + 75 % P + PSB (Rs. 60920 and 58680/ha). Total tuber number during first year was higher with recommended dose of NK + 75% P + phosphobacteria (439.0 thousand/ha) followed by recommended doses of NK + 50% P + PSB (436.0 thousand/ha). Whereas, in 2006-07, total tuber number were higher in the treatment with recommended dose of NPK (427.0 thousand/ha) followed by recommended doses of NK + 75% P + PSB (389.5 thousand/ha).

At Deesa, there was no significant difference among various treatments during both the years. However, in comparison to other treatments, recommended dose of N K + PSB gave highest total tuber yield (42.15 and 39.29t/ha). Highest net returns of Rs.146336 and 164071/ha were obtained with the use of recommended dose of NK + PSB followed by recommended NK + 50% P + PSB and only PSB (Rs. 143255 and 157451/ha).

At Modipuram, there was no significant difference among various treatments during both the years. Results showed that the effect of treatments was non-significant for total yield and numbers of tubers (Table 48, 49 and 50). However, the maximum tuber yield (29.0 and 30.16 t/ha) and number (733.2 and 742.5 thousand/ha) were obtained with recommended NPK. Highest net gain (Rs. 34723 and 37498/ha) was also obtained in the same treatment followed by recommended dose of N K (Rs. 33348 and 37516/ha).

On the basis of mean value of two years amongst all the treatments, Maximum net returns (Rs 98837 63632, 35303, 67397 and 36111/ha) were obtained with recommended dose of NPK closely followed by recommended dose of NK + 75% P + phosphobacteria (Rs 98271, 58134, 28260, 59800 and 32364/ha) at Chhindwara, Kota, Dholi, Hisar and Modipuram, respectively (Table 51 and Fig.19). Whereas, at Kalyani and Hassan, application of recommended dose of NK + 75% P + phosphobacteria gave maximum net return followed by recommended dose of NPK. However, at Deesa, recommended dose of NK + phosphobacteria gave maximum net return (Rs 151894/ha) followed by recommended dose of NPK (Rs 145814/ha). Therefore, the net return indicated that 25 to 50% dose of P can be saved by the PSB treatment.

The tuber yield obtained in presence of phosphobacteria with respect to varying P levels from 0-75% fitted well in quadratic model with R^2 value ranging from 0.713 for Bhubaneshwar to 0.998 for Ooty. Taking tuber yield at recommended NPK as target yield when % of recommended P required in presence of PSB was calculated, Hassan Center showed maximum saving. At this center only 57% of recommended P application in presence of PSB, achieved the yield what recommended NPK gave and therefore, PSB treatment saved about 43% of P fertilizer. In the per cent saving of P fertilizer due to PSB use Hassan was followed by Kalyani (24%), Chhindwara (20%), Ooty (19%), Patna (16%), and Bhubaneshwar (11%). At Kota and Dholi, PSB treatment could save only a meager amount (*i.e.* 6 and 3%, respectively) of recommended P (Table 51). Therefore, it may be inferred that effectiveness of PSB was dependent on the agro-ecological situation and soil conditions of the field where it is being used.

Table 48: Effect of PSB and graded dose of P on total tuber yield (t/ha) at different centers

Treatments	Center											
	Kalyani			Chhindwara			Kota			Patna		
	I	II	Pooled	I	II	Pooled	I	II	Pooled	I	II	Pooled
T1	22.6	22.17	22.39	25.88	27.15	26.52	14.69	14.22	14.46	30.35	26.68	28.52
T2	26.71	27.19	26.95	31.14	32.54	31.84	27.43	22	24.72	35.86	30.66	33.26
T3	26.39	25.38	25.89	30.2	31.05	30.63	26.38	20.77	23.58	34.73	29.65	32.19
T4	25.64	25.64	25.64	27.39	29.07	28.23	22.41	15.11	18.76	32.83	29.41	31.12
T5	24.15	23.51	23.83	26.5	28.52	27.51	17.24	15.11	16.18	31.11	27.73	29.42
T6	27.09	26.61	26.85	31.63	33.05	32.34	28.56	23.97	26.27	36	30.87	33.44
CD (0.05)	NS	3.7	3.79	1.76	2.74	3.08	1.05	1.52	1.39	1.7	3.05	2.36
	Hassan			Ooty			Bhubaneshwar			Dholi		
	I	II	Pooled	I	II	Pooled	I	II	Pooled	I	II	Pooled
T1	24.65	19.53	22.09	9.49	4.79	7.14	8.32	13.15	10.74	11.39	6.79	9.09
T2	44.15	28.23	36.19	21.3	9.82	15.56	9.47	20.58	15.03	16.02	10.34	13.18
T3	34.71	25.79	30.25	19.15	6.32	12.74	4.99	17.56	11.28	14.09	9.57	11.83
T4	31.94	22.81	27.38	14.38	5.95	10.17	8.41	16.76	12.59	13.12	8.95	11.04
T5	28.38	16.15	22.27	12.32	5.31	8.815	5.15	14.66	9.905	12.16	7.26	9.71
T6	37.71	26.98	32.35	21.78	11.15	16.47	9.27	22.51	15.89	16.41	11.81	14.11
CD (0.05)	4.52	1.46	3.22	6.68	3.38	5.12	NS	1.39	3.22	2.78	0.99	1.99
	Hisar			Deesa			*Modipuram*					
	I	II	Pooled	I	II	Pooled	I	II	Pooled			
T1	25.92	25.78	25.85	40.67	39.26	39.97	28.01	29.57	28.79			
T2	29.51	28.95	29.23	39.85	37.72	38.79	27.4	28.99	28.2			
T3	28.97	28.45	28.71	41.74	36.17	38.96	26.98	28.47	27.73			
T4	27.7	27.41	27.56	39.16	38.07	38.62	27.74	29.17	28.46			
T5	26.94	26.11	26.53	42.15	38.19	40.17	26.95	28.65	27.8			
T6	30.55	31.65	31.1	40.56	38.09	39.33	29.01	30.16	29.59			
CD (0.05)	2.56	5.82	4.30	NS	NS	NS	NS	2.9	NS			

T1: Recommended dose of N K, T2: Recommended dose of NK + 75% P + phosphobacteria, T3: Recommended dose of NK + 50% P + phosphobacteria, T4: Recommended dose of NK + 25% P + phosphobacteria, T5: Recommended dose of NK + phosphobacteria and T6: Recommended dose of NPK

Table 49: Effect of PSB and graded dose of P on total tuber number ('000/ha) at different centers

Treatments	Center														
		Kalyani			Chhindwara			Kota			Patna			Hassan	
	I	II	Pooled	I	II	Pooled	I	II	Pooled	I	II	Pooled	I	II	Pooled
T1	520.8	505	513.1	578.1	603.4	590.8	755.4	908.2	831.8	692.5	682.3	687.4	257.3	219.4	238.4
T2	580.7	580	580.4	536.5	557.9	547.2	672.3	508.5	590.4	869.3	779.2	824.25	388.5	359.4	374
T3	543.8	521	532.3	567.3	581	574.2	782.4	601.1	691.8	782.5	716.7	749.6	261.7	332.7	297.2
T4	598.8	601	599.9	485	513.1	499.1	776.7	620.6	698.7	748.4	771	759.7	249.3	280.2	264.8
T5	543.8	550	546.8	470.7	500.8	485.8	611.4	656.8	634.1	691.8	717.7	704.75	251.4	183.6	217.5
T6	589.2	613	601.2	552.5	580.2	566.4	568.4	454.7	511.6	783.7	736.6	760.15	255.2	340.3	297.8
CD (0.05)	NS	NS	NS	37.3	44.2	39.2	NS	71.9	120.4	81.7	NS	91.5	62.6	19.8	13.21
		Ooty		Bhubaneshwar		Dholi			Hisar		Deesa		Modipuram		
	I	II	Pooled	II	I	II	Pooled	I	II	Pooled	II	I	II	Pooled	
T1	450.5	263	356.8	293.8	275.7	268.1	271.9	356.3	353.4	354.85	644.6	715.9	728.8	722.4	
T2	858.1	412	634.8	453	378.5	345.1	361.8	439	389.5	414.25	652.3	725.7	738.2	732	
T3	668	305	486.4	409	353	317.5	335.3	436	367.3	401.65	216.7	688.5	715.7	702.1	
T4	692.7	292	492.2	379.6	311.3	292.1	301.7	410.1	361.9	386	727.5	715.9	725.9	720.9	
T5	467.4	294	380.9	340	294	280.9	287.5	412.2	349	380.6	583.3	710.6	724.9	717.8	
T6	815.1	490	652.4	500.2	402.4	394.9	398.7	421.7	426.5	424.1	709.2	733.2	742.5	737.9	
CD (0.05)	179.4	132	151.1	21.7	81.5	44.8	63.08	46.5	NS	59.56	NS	NS	NS	NS	

T1: Recommended dose of N K, T2: Recommended dose of NK + 75% P + phosphobacteria, T3: Recommended dose of NK + 50% P + phosphobacteria, T4: Recommended dose of NK + 25% P + phosphobacteria, T5: Recommended dose of NK + phosphobacteria and T6: Recommended dose of NPK

Table 50 : Effect of PSB and graded dose of P on net return (Rs/ha) at different centers

Treatments	Centers											
	Kalyani		**Chhindwara**		**Kota**		**Patna**		**Hassan**		**Ooty**	
	I	**II**	**I**	**II**	**I**	**II**	**I**	**II**	**I**	**II**	**I**	**II**
T1	34901	21569	63599	92839	19835	17963.88	47336	-25620	64980	47700	-25620	NR
T2	44639	33300	80784	115757	69129	47138.88	62673	25513	107925	87725	25513	*NR*
T3	44526	28723	77870	109429	65458	42823.90	59636	17459	94540	72550	17459	*NR*
T4	41614	29079	68452	100901	49629	20855.90	54147	-4267	88975	57575	-4267	*NR*
T5	38701	25057	65884	98520	29993	21481.88	49383	-12884	54940	30000	-12884	*NR*
T6	46551	29870	82103	115571	731610	54103.88	63022	26997	99840	85000	26997	*NR*
	Bhubaneshwar		**Dholi**		**Hisar**		**Deesa**		**Modipuram**			
	I	**II**	**I**	**II**	**I**	**II**	**I**	**II**	**I**	**II**		
T1	NR	- 6432	20800	4365	47943	47399	139141	164071	33348	37516		
T2	*NR*	21850	36250	20270	60920	58680	133342	153122	30293	34435		
T3	*NR*	10238	33850	16275	59081	56997	143255	144325	29463	33742		
T4	*NR*	7499	27600	14205	54310	53154	130695	156225	32143	35899		
T5	*NR*	- 424	24250	8180	51631	48311	146336	157451	30143	34761		
T6	*NR*	29142	40950	29655	65159	69635	136589	155039	34723	37498		

T1: Recommended dose of N K, T2: Recommended dose of NK + 75% P + phosphobacteria, T3: Recommended dose of NK + 50% P + phosphobacteria, T4: Recommended dose of NK + 25% P + phosphobacteria, T5: Recommended dose of NK + phosphobacteria and T6: Recommended dose of NPK

Table 50: Effect of PSB and graded dose of P on net return (Rs/ha) at different centers (mean of two years)

	Kalyani	Chhindwara	Kota	Hassan	Dholi	Hisar	Deesa	Modipuram
T1	28235	78219	18899	56340	12583	47671	151606	35432
T2	38970	98271	58134	97825	28260	59800	143232	32364
T3	36625	93650	54141	83545	25063	58039	143790	31603
T4	35347	84677	35242	73275	20903	53732	143460	34021
T5	31879	82202	25737	42470	16215	49971	151894	32452
T6	38211	98837	63632	92420	35303	67397	145814	36111

T1: Recommended dose of N K, T2: Recommended dose of NK + 75% P + phosphobacteria, T3: Recommended dose of NK + 50% P + phosphobacteria, T4: Recommended dose of NK + 25% P + phosphobacteria, T5: Recommended dose of NK + phosphobacteria and T6: Recommended dose of NPK

Table 51: Contribution of PSB in phosphorus nutrition of potato crop as derived from quadratic ($y= ax^2+bx+c$) model

Quadratic coefficients	Kalyani	Chhindwara	Kota	Patna	Hassan	Ooty	Bhubaneshwar	Dholi
a	-0.30	0.196	-0.576	-0.252	0.332	0.586	0.426	0.008
b	60.94	46.86	164.9	69.26	153.6	47.27	24.31	44.2
c	23949	27367	15884	29452	22536	8766	10357.75	9765
R^2	0.944	0.966	0.963	0.997	0.986	0.998	0.713	0.990
Target Yield* (kg/ha)	26850	32340	26270	33440	32350	16470	15890	14110
Dose of P required in presence of PSB (%)	76	80	94	82	57	81	89	97
Contribution of PSB (%)	24	20	6	18	43	19	11	3

*Yield obtained at recommended dose of fertilizer

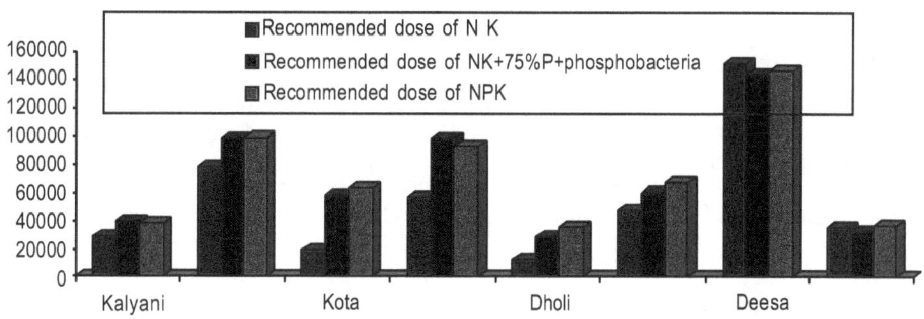

Fig. 19: Effect of PSB and graded dose of P on average net return (Rs/ha) of two year at different centers

The results of this study indicated the beneficial effects of PSB inoculation to potato at most of the centers. It was evident that at most places PSB inoculation was capable of saving at least 25% of recommended phosphatic fertilizers as the yield obtained from PSB inoculation replacing 25% of recommended P was superior or at par with recommended NPK at centers Kalyani, Chhindwara, Kota, Patna, Hassan and Ooty. Centers like Bhubaneshwar, Dholi, Hisar, Deesa and Modipuram did not show consistency in P saving with the use of PSB. At Dholi and Hisar, soils were low in organic matter (< 0.5%) and high in pH (> 8.0) which was not very favourable for the PSB group of microbes. However, some response was due to P solubilised possibly from inorganic insoluble calcium compounds. The poor response at Modipuram was possibly resulting from high available P in soil and at Deesa the very low organic carbon (<0.2 %) might not have allowed PSB to proliferate and solubilise the unavailable P. The beneficial effect and saving of P fertilizer using PSB was reported by Singh (2005) and Singh, (2002) in north-eastern hills and under East Khasi hill conditions of Meghalaya, Sood, and Sharma (2001) at Shimla hills and Sud, and Jatav (2007) in brown hill soils of Shimla (HP)). Phosphate solubilizing bacteria (PSB) are capable of hydrolyzing organic and inorganic phosphorus from insoluble compounds. It is generally accepted that the mechanism of mineral phosphate solubilization by PSB strains is associated with the release of low molecular weight organic acids, through which their hydroxyl and carboxyl groups chelate the cations bound to phosphate, thereby, converting it into soluble forms. In addition, some PSB produce phosphatase like phytase that hydrolyse organic forms of phosphate compounds efficiently (Zehra, 2010; Rodriguez and Fraga, 1999). At Chhindwara replacing of 25% of inorganic P with PSB without compromising on yield despite low available soil P was possibly due to the capability of these microbes to mobilize this nutrient from organic sources as soil of the experimental field was having around 0.7% organic carbon. Similarly, 0.6 % organic carbon in the experimental soils at Kota might have helped the microbes. In addition, pH above 7.5 indicates that P solubilization in these soils might have come from Ca-bound compounds. Even at very low pH (4.0-4.2) good response of PSB at Ooty was possible only due to high content of organic carbon (1.60-1.64 %) in soil. It is most likely that major fraction of P solubilised by PSB had come from organic sources.

The results showed that phosphate solubilizing bacteria (PSB) is of great importance in reducing fertilizer doses and practicing integrated plant nutrients supply. It may be concluded that combined application of P along with tuber inoculation with PSB can be effective in reducing the inorganic P dose by approximately 20 to 25%. Besides saving fertilizer, PSB also showed increased efficiency, net return and B: C ratio. These information have special importance in the light of wide spread finding that many of the potato growing pockets are showing build up of P in soil creating imbalances of nutrients particularly making some micronutrient unavailable (Jatav and Sud, 2007; Jatav *et al.*, 2011, 2012, 2015 and 2016).

13

Soil Test Based Phosphorus Recommendations in Different Potato Growing Regions

Maximum potato yield occurs when sufficient P is available during early vegetative development and the entire period of tuber growth. Total plant P uptake increases rapidly during tuber initiation, levels off to a constant rate during tuber bulking, and ceases with plant maturation occurs primarily through the transfer of P reserves from the vine and roots. Phosphorus uptake by potatoes is relatively low compared with uptake of potassium or nitrogen but similar to uptake of sulfur. The amount of P in the soil solution that is readily available for plant uptake is very small compared with the total amount of P in the soil.

A general blanket application of fertilizer leads to its poor efficacy and wastage of costly resources as well as of environmental pollution. This becomes more relevant in case of P as unlike nitrogen applied P remains in soil and adds to its build up. In addition, this has resulted in severe imbalance of nutrients in soil (Kumar *et al.*, 2009).

Keeping this in view, a multilocation experiment was conducted in the major potato growing states of India under the All India Coordinated Research Project on Potato to evaluate suitable dose of P based on soil test on P economy in potato (table 52). Exact quantification of phosphorus fertilizer based on soil test is very important.

Table 52: Treatments details for different locations of AICRP (Potato) centers based on soil test

T1	No phosphorus application	
T2	Application of recommended dose of P	
T3*	Application of P as per soil test value	
T4*	Application of P as per soil test value	
Soil test values: (Olsen's P)	T3*	T4*
< 10 ppm	75% of recommended P	125% of recommended P
10-20 ppm	25% of recommended P	50% of recommended P
>20 ppm	12.5% of recommended P	25% of recommended P

Potato tuber yield from soil having low P availability (<10 ppm)

Soils of experimentation field at Bhubaneshwar (Orissa), Chhindwara (MP), Dholi (Bihar), Hisar (Haryana), Jalandhar (Punjab), Ooty (Tamilnadu), Kalyani (West Bengal) and Kota (Rajasthan) had available P less than 10 ppm, therefore 75% of recommended P and 125% of recommended P were applied as T3 and T4 respectively (table 53, 54 and 55).

At Chhindwara, during both the year, total tuber yield (28 and 26.12 t/ha) and net return (Rs. 71580 and 88740/ha) were highest in treatment receiving 125% of recommended P applied followed by application of recommended dose of P (27.2 and 25.51 t/ha and Rs 69191 and 86455/ha), respectively. Highest total number of tubers was also obtained with 125% of recommended P application followed by application of recommended dose of P. During first and second years of experimentation the percent increase in yield was 2.9 and 2.4% and net return was 3.5 and 2.6%, respectively in the treatment receiving 125% recommended dose of P over 100% recommended dose of P. During both the year, application 125% of recommended P gave maximum numbers (447.3 and 442.9 thousand /ha,) followed by recommended dose of P (439.0 and 417.6 thousand /ha,).

At Dholi, the trial was conducted on soil with available P 18 kg/ha. Higher tuber yield (13.09 t/ha) and net return (Rs. 22450/ha) were obtained by the application of 125% of recommended P. The yield at recommended dose of P was though at par, the net return was Rs 4300/ha less as compare to 125% recommended dose of P.

The availability of P in the experimental plot before planting at Hisar was 22 kg P/ha (< 10 ppm). Total tuber yield and total number of tubers was highest in recommended dose of P followed by 125% of recommended P. However, the difference was not significant. These parameters were lowest where P was not applied. Net gain was highest in the treatment of recommended dose of P (Rs. 37025/ha) and lowest without P application (Rs. 26365/ha). Which was Rs 1538/ ha higher than net return at 125% of recommended P application.

At Jalandhar, recommended P application recorded higher values for total tuber yield (29.6 t/ha) and net returns (Rs. 39750/ha) followed by 125% of recommended P (28.4 t/ha and Rs. 35980/ha, respectively). The least values for these two attributes were recorded without P application (23.5 t/ha and Rs. 22150/ha, respectively). During second year, 125% recommended P application recorded higher values for total tuber yield (29.06 t/ha) and net returns (Rs. 38080/ha) followed by application of recommended dose of P (28.71 t/ha) yield and Rs.37050/ha net return). The total number of tubers were highest with application of recommended dose of P (477.4 thousand/ha) followed by 75% of recommended P (447.2 thousand /ha). The tuber yield however, was statistically at par with recommended P and 125% of recommended P during both the year.

At Ooty, during both the year, tuber yield (34.73 t/ha), net returns (Rs.84613/ha) and total tuber numbers (690.1 thousand/ha) were higher with 125% recommended dose of P followed by application of recommended dose P. During second year, yield followed similar trend but the number of tubers were not influenced by the treatments.

At Kalyani, during both the year, the differences among treatments for yield were also non-significant except for total number of tubers which was highest at 75% of recommended P. Highest net return (Rs. 52851/ha) was recorded for the recommended dose of P. During second year, the highest yield of tubers (28.53 t/ha) and net return (Rs. 36586/ha) was recorded with 125% of recommended P followed by application of recommended dose of P (28.10 t/ha tuber yield and Rs.35639/ha net returns). These two treatments were statistically at par. Total tuber numbers (597.2 thousand /ha) were higher in treatment where 75% of recommended P applied.

At Kota, highest total yield (23.89 t/ha) and net returns (Rs. 55345/ha) were obtained in treatment receiving 125% of recommended P which was 4.4 and 6.9% higher as compared to application of recommended dose of P. Highest numbers of total tubers (688 and 642 thousand /ha) were harvested in treatment where P was not applied. During second year, the highest total yield (20.11 t/ha) was obtained in treatment receiving application of recommended dose of P was applied followed by 125% of recommended P (20.03 t/ha). The highest net returns (Rs. 38980/ha) were obtained with application of recommended dose of P followed by 125% of recommended P (Rs. 38047/ha). At Bhubaneshwar, highest tuber yield (15.19 t/ha) was obtained with 125% of recommended P application which was significantly better than any other treatments and this treatment also gave the highest net return. Under low P conditions, application of recommended dose remains insufficient and optimum yield was not achieved at all the center. At these centers when 100-125% of recommended P was applied, yield improved. This result confirms the recommendation that at deficient soils application of P may be increased by 25% above recommended dose (Ess, *et al*, 2003; Goswami, 2006; Goswami, *et al*, 2009; Sud, *et al*, 2008 and Trehan *et al*, 2008).

Potato tuber yield from soil having medium P availability (10-20 ppm)

Soils of experimental field at location namely Patna (Bihar), Pantnagar (Uttrakhand) and Bhubaneshwar (Orissa), had available P (Olsons, P) between 10-20 ppm, therefore 25% of recommended P and 50% of recommended P were taken T3 and T4, respectively (table 56, 57 and 58).

At Patna, recommended P gave highest tuber yield (36.87 t/ha) and the net return (Rs. 65535/ha) followed 50% of recommended P. During second year also, the higher tuber yield (24.98 t/ha) was obtained at recommended dose of P, which was statistically superior to any other treatments. The net return was also highest with recommended dose of P (Rs. 48349/ha) followed 50% of recommended P (Rs. 39818/ha).

At Pantnagar, the highest total yield (34.21 t/ha) were recorded in treatments where application of recommended dose of P applied which was significantly better than any other treatment. The 50% of recommended and recommended dose of P produced highest number of total tubers (487.1 and 459.7 thousand/ha, respectively). The net return was also higher in treatment receiving 50% of recommended P (Rs. 36219/ha) followed by recommended dose of P (Rs 34882/ha) as compared to other treatments. During first and second years, yield increased by 7.08 and 10.09% and net return increased by 11.4 and 21.042%, respectively at recommended dose of P as compare to 50% P application.

At Bhubaneshwar, application of recommended doses of P gave the highest yield (22.54 t/ha) which was significantly better than any other treatments. The treatment not receiving any P resulted in lowerest yield. Recommended dose of P also produced significantly highest number of tubers 460.1 thousand/ha. In terms of net returns, recommended dose of P produced highest return (Rs.29154/ha) followed by 50% of recommended dose of P (Rs. 22699/ha). Therefore, at these centers, as available P was in medium range, the recommended P application gets the best yield. However, statistically at many places 50% of recommended P did not decrease the yield significantly and was at par with recommended P (Ess, *et al.*, 2003; Goswami, 2006; Goswami, *et al.*, 2009; Sud, *et al.*, 2008 and Trehan *et al.*, 2008).

Potato tuber yield from soil having high P availability (>20 ppm)

Soils of experimentation field at Deesa (Gujrat) had available P (Olsons, P) more than 10 ppm, therefore 12.5% of recommended P and 25% of recommended P (T3 and T4) were taken. Results revealed non-significant differences among the treatments for tubers yield. Highest tubers yield (46.21 t/ha) was obtained at recommended dose of P followed by the treatment receiving 25% of recommended P (45.69 t/ha, whereas, maximum number of tubers were recorded in treatment receiving recommended dose P (601944/ha) followed by the treatment not receiving any P (590139/ha). Treatment that received 25% of the recommended dose of P recorded highest net return (Rs.163749/ha) followed by the recommended dose of P application (Rs. 162836/ha).

Table 53 : Total yield (t/ha) in the soils having < 10ppm initial P at different locations

Treatments	Bhubaneshwar	Chhindwara		Dholi	Hisar	Jalandhar		Ooty		Kalyani		Kota	
	2005-06	2005-06	2006-07	2005-06	2005-06	2005-06	2006-07	2006	2007	2005-06	2006-07	2005-06	2006-07
T1	12.26	24.41	19.5	8.24	20.54	23.49	25.04	29.88	12.95	24.63	24.09	13.57	12.2
T2	11.52	27.24	25.51	12.12	23.53	29.55	28.71	31.77	16.13	29.22	28.1	22.89	20.11
T3	10.4	26.18	22.16	10.56	21.56	27.24	27.67	30.98	14.54	27.35	27.46	20.75	16.37
T4	15.19	28.04	26.12	13.09	22.08	28.35	29.06	34.73	17.58	28.42	28.53	23.89	20.03
SEd	1.1	0.78	0.81	1.37	0.65	1.82	0.57	6.15	2.88	3.01	1.82	0.67	0.77
CD (0.05)	2.49	1.76	1.83	3.09	1.46	4.11	1.28	13.9	6.51	NS	4.12	1.52	1.74
CV %	12.6	4.16	4.9	17.57	4.17	9.48	2.9	27.3	26.59	15.53	9.53	4.7	6.33

T1: No phosphorus application, T: Application of recommended dose of P, T3: 75% of recommended P and T4: 125% of recommended P

Table 54 : Total tuber number in the soils having < 10ppm initial P at different locations (thousand/ha).

Treatments	Bhubaneshwar	Chhindwara		Dholi	Hisar	Jalandhar		Ooty		Kalyani		Kota	
	2005-06	2005-06	2006-07	2005-06	2005-06	2005-06	2006-07	2006	2007	2005-06	2006-07	2005-06	2006-07
T1	400.3	389.7	337.2	244.4	280.5	303.1	418.9	575.5	450.5	492.0	453.5	688.9	642.6
T2	330.1	439.0	417.6	239.6	317.5	396.5	477.4	622.4	535.2	580.7	596.7	618.4	566.7
T3	369.4	415.5	367.7	228.0	293.2	401.9	447.2	600.3	498.7	611.6	597.2	606.3	493.6
T4	399.1	447.3	442.9	271.4	311.2	402.3	439.4	690.1	574.2	552.4	529.9	500.6	424.2
SEd	29.0	14.2	19.6	62.7	21.0	16.2	14.7	34.4	74.7	35.5	34.9	40.2	22.2
CD (0.05)	330	439	418	240	318	397	477	622	535	581	597	618	567
CV %	10.95	4.76	7.1	36.04	9.88	6.11	4.68	7.83	20.53	8.98	9.08	9.43	5.9

T1: No phosphorus application, T: Application of recommended dose of P, T3: 75% of recommended P and T4: 125% of recommended P

Table 55 : Net return (Rs/ha) in the soils having < 10ppm initial P at different locations

Treatments	Bhubaneshwar	Chhindwara		Dholi	Hisar	Jalandhar		Ooty		Kalyani		Kota	
	2005-06	2005-06	2006-07	2005-06	2005-06	2005-06	2006-07	2006	2007	2005-06	2006-07	2005-06	2006-07
T1	-3772	60886	61205	5050	26365	22150	26950	48576	9418	40851	26969	15335	9844
T2	-8590	69191	86455	22450	37025	39750	37050	71980	25756	52851	35639	50473	38980
T3	-11649	65867	71802	15100	30128	33030	34230	69113	16942	47839	34654	42434	24265
T4	7015	71580	88740	26750	31887	35980	38080	84613	33793	50064	36586	53963	38047

T1: No phosphorus application, T: Application of recommended dose of P, T3: 75% of recommended P and T4: 125% of recommended P

Table 56: Total yield (t/ha) in the soils having 10-20 ppm and > 20ppm P at different locations

Treatments	Soils having 10-20 ppm*				Soils having > 20ppm **
	Patna		Pantnagar	Bhubaneshwar	Deesa
	2005-06	2006-07	2006-07	2006-07	2005-06
T1	27.94	19.97	23.47	17.09	43.83
T2	36.87	24.98	34.21	22.54	46.21
T3	30.95	22.35	28.47	18.75	42.23
T4	34.43	22.69	32.37	20.69	45.69
SEd	2.82	0.94	0.67	0.61	3.35
CD (0.05)	6.36	2.1	1.51	1.38	NS
CV %	12.24	5.92	3.18	4.38	13.04

*T1: No phosphorus application, T: Application of recommended dose of P, T3: 25% of recommended P and T4: 50% of recommended P for soils having 10-20 ppm

**T1: No phosphorus application, T: Application of recommended dose of P, T3: 12.5% of recommended P and T4: 25% of recommended P for soils having e”20 ppm

Since the soil was high in available P, no significantly effect of P application was observed. This is due to build up of available P has reached a level where response of its application may not be obtained in next few coming crops. However, minimum maintains dose of P may be applied to avoid the risk of yield loss in subsequent years, P level may get depleted and need to be supplemented with recommended dose based on soil test (Ess, *et al.*, 2003; Goswami, 2006; Goswami, *et al.*, 2009; Sud, *et al.*, 2008 and Trehan *et al.*, 2008).

Table 57: Total tuber number in the soils having 10-20 ppm and > 20 ppm P at different locations

Treatments	**Soils having 10-20 ppm**				**> 20ppm Deesa**
	Patna		**Pantnagar**	**Bhubaneshwar**	**Deesa**
	2005-06	**2006-07**	**2006-07**	**2006-07**	**2005-06**
T1	643.2	595.0	429.4	330.8	590.1
T2	796.7	630.9	459.7	460.5	601.9
T3	729.8	584.9	422.5	359.4	560.3
T4	725.1	637.3	487.1	439.4	580.7
SEd	61.3	47.6	10.7	8.4	33.9
CD (0.05)	NS	NS	24275	18935	NS
CV %	11.97	11	3.38	2.98	10.07

*T1: No phosphorus application, T: Application of recommended dose of P, T3: 25% of recommended P and T4: 50% of recommended P for soils having 10-20 ppm

**T1: No phosphorus application, T: Application of recommended dose of P, T3: 12.5% of recommended P and T4: 25% of recommended P for soils having >20 ppm

Table 58: Net return (Rs/ha) in the soils having 10-20 ppm and > 20 ppm P at different locations

Treatments	**Soils having 10-20 ppm***				**Soils having > 20ppm**
	Patna		**Pantnagar**	**Bhubaneshwar**	**Deesa**
	2005-06	**2006-07**	**2006-07**	**2006-07**	**2005-06**
T1	40011	29553	14654.88	9252	154961
T2	65535	48349	34882.98	29154	162836
T3	48717	38648	27087.28	15399	145905
T4	58827	39818	36218.58	22699	163749

*T1: No phosphorus application, T: Application of recommended dose of P, T3: 25% of recommended P and T4: 50% of recommended P for soils having 10-20 ppm

**T1: No phosphorus application, T: Application of recommended dose of P, T3: 12.5% of recommended P and T4: 25% of recommended P for soils having >20 ppm

Study to economize on P dose from fertilizer in different P status of soils in different location revealed that soil having low P availability (<10 ppm) responded to P fertilizers and there is need of P application. Thirteen experiments conducted at low P status soils (<10 ppm), at different centers 100-125% recommended dose of P gave best tuber yield and also net returns. However, in case of medium P availability soils (10 to 22 ppm), the recommended dose of P fertilizer was found superior compared to other treatments. Four experiments conducted at medium P status soils (10-20 ppm) 50% recommended dose of P or recommended dose of P was sufficient to meet P need of the crop and gave higher yield and also net returns. In case of high P status soils (>20 ppm), application of P did not respond. Therefore, finding clearly outlined the importance of soil test in efficiently use of costly P fertilizer and economize the cost of cultivation in soils testing high in P (Jatav *et al.*, 2013).

14

Fertilizers Recommendations using GIS and Remote Sensing in the Intensively Growing Potato Regions of Punjab

Potato plays a vital role in food security for ever increasing world population (Thiele *et al.*, 2010; Scott and Sourez, 2011; Scott and Sourez, 2012). It is highly capital and labour intensive crop (Kushwah and Singh, 2011). Presently, India is the second largest potato producer in the world. India is producing about 45-48 million tonnes of potato annually from about 2 million hectare of area. However, India would require 125 million metric tonnes of potato from 3.6 million hectare area in 2050 (Vision, 2050). The stagnating growth rates of cereals' productivity, large scale diversion of food grains to feed & bio-fuel and expected steep rise in per capita consumption of pulses, edible oil, fruits, vegetables, milk, sugar and non-vegetarian food in the regime of steadily rising population is bound to put pressure on existing cultivable land. Since, cultivable land is expected to remain more or less constant in the next 40 years, the role of crops like potato having higher production potential per unit land and time will become imperative. In this context potato crop has very high probability of making crucial contribution to the future national food security agenda. The perceived changes in Indian socio-economics in the medium and long term are expected to enhance per capita food consumption of fresh potatoes

Fertilizer and nutrient recommendations for different crops are mostly region specific and very general in nature which results into a blanket application of fertilizer leading to its poor efficacy and wastage of costly resources as well as of environmental pollution. In addition, this has caused severe imbalance of nutrients in soil. The spatial variability of available nutrients properly mapped along with other relevant parameters will help to develop site specific nutrient management practices with increased productivity and profitability apart from checking nutrient imbalance and environmental pollution. The proper understanding of spatial nutrient variability, if integrated with fertilizer recommendation for the cropping system, can ensure the fertilizer use in adequate and balanced manner (Sen *et al.*, 2008). Since, the rate of nutrient application will match with local available nutrient status, there will be improvement in efficiency and productivity with minimum damage to the environment.

Punjab being an important potato (*Solanum tuberosum* L.) growing state of India and contributes about 5% to national acreage and production. In Punjab, Jalandhar is the number one district with about 25% of potato acreage and production shares. In Jalandhar district there are few pockets where majority of farmers follow potato based cropping systems and during the main season a very large continous area of potato crop can be seen. Potato being the most important and responsive crop in the system receives maximum share of fertilizers. In addition the potato cultivation as well as other agricultural practices are highly mechanized and farmers generally have very small and meager amount of farm or animal based organic waste to recycle to the fields. Besides, farmers in rice-potato system don't allow rice crop residue to decompose and it is a general practice to burn the residue to facilitate early or timely planting of potato. These compulsive situations have led to severe deficiency and imbalances of nutrients in these potato growing pockets. Imbalances of available nutrients particularly build up of P and deficiency of K at many locations has been reported (Sharma 2004, Trehan *et al.*, 2008). These high intensity cropping, poor recycling of organic residue and use of heavy doses of micronutrient free high analysis fertilizers has given rise to wide spread deficiency of even micronutrients in different districts of Punjab (Bansal *et al.,* 1986, Singh *et al.,* 1988). There seems urgent need to spatially map the emerging nutrient imbalances and deficiencies to suggest pocket specific management practices for these nutrients. Use of GPS in identifying exact location of sampling and further using GIS software for development of spatial distribution map can be very important in recommending the site specific nutrient management. Therefore, the present study was undertaken to assess the major (N, P and K) and micro (Fe, Cu, Zn and Mn) nutrient status of soils of potato growing areas in Jalandhar district of Punjab and to map their spatial variability using GPS and GIS to refine fertilizer recommendations and promote site specific nutrient management.

The potato growing pockets of Jalandhar district were demarcated based on the available information, ground truth and classification of the IRS P6 AWiFS image (Dua *et al.,* 2007). Approximate sampling points were marked in the classified image by counting number of pixels under potato crop to have atleast one sample point in every one to two square kilometers of potato area. Actual sampling was done while transacting the area through different roads and attempting to reach very close to the geographical coordinates of approximate sampling points mark in the image with the help of GPS. Representative soil samples of potato fields from 0 to 0.15 m depth were collected during October 2009 at every one to two kilometer distance using GPS receiver (E-TREX-VISTA, Garmin) to record geographical coordinates. To account for greater variability, depending upon visual observation, more number of samples were collected wherever needed. Sampling intensity was kept more in pockets having larger proportion of potato crop. Collected soil samples were processed and analysed for their chemical properties using standard techniques. The actual soil sample points marked using GPS were fed into the GIS environment in already digitized boundary of Jalandhar district. The classified potato area in IRS P6 AWiFS image was also taken as background to generate the sampling site map. Values of different parameters (like pH, OC, available P and K and micronutrients) were tagged with corresponding points and interpolation maps for each individual parameters were prepared using suitable semivariogram and kriging in remote sensing GIS software "Geomatica". Further, the maps generated for pH, OC, P, K and micronutrients (Zn, Cu, Mn and Fe) were classified taking the ranges for different parameters suitable for potato cultivation. The integration of classified maps of P and K and that of four micronutrients were generated to find out specific locale of single or multinutrients deficiency. GIS software was also used to estimate the area falling under different class of available nutrients.

Sampling points: Out of 205 approximate sampling points marked in the image 35 points were sampled with less than 20 m of accuracy based on GPS navigation, 72 points were sampled with 20 to 50 m accuracy and 88 points were sampled with 50 to 100m accuracy. Remaining 10 points could not be sampled as we could not reach within 100m of approximate points marked in image due to some obstruction or absence of approach road.

Spatial distribution of available major and micronutrients: Spatial variability maps were prepared after kriging of point values of different parameters in 'Geomatica' environment. Classification results clearly showed the specific locales of the pockets where attention is required with respect to management of major and micronutrients. *Soil pH and organic carbon:* After kriging for soil pH and classification in the five classes in the range of 6.0 to 6.5, 6.5 to 7.0 and 7.0 to 7.5, 7.5 to 8.0 and more than 8.0 potato area falling under these categories were 0.7, 15.5, 36.2, 29.0 and 18.8%, respectively (Table 1 and Fig 1). The 47% area plunged in last two categories with pH value more than 7.5 may be considered a situation not very suitable for potato cultivation. This figure is somewhat smaller when compared with number of samples (55%) above pH 7.5. The neutral to slightly alkaline pH may be attributed to application of higher amount of inorganic fertilizer material in the soil for potato crop, which resulted in the retention of basic cations on the exchangeable complex of the soil and low level of organic carbon which resulted in poor buffering capacity (Sharma *et al.*, 2008). Similarly, area under organic carbon in range <0.2, 0.2 to

0.3, 0.3 to 0.4, 0.4 to 0.5 and >0.5%, were 8.6, 44.0, 37.9, 9.3 and 0.3%, respectively (Table 1 and Fig 1). In fact, this situation of higher pH value (near 8) with low organic carbon cannot be treated as most suitable situation for potato cultivation. The area having organic carbon in medium range was much smaller (not even 0.5%) as compared to number of samples (2.6%). This happened because while kriging value at any point depended on weight given to value of different nearby samples. There were more points with lower value, so proportionally more area came under lower organic carbon.

Available phosphorus and potassium: Area under medium (10-20 ppm), high (20-30 ppm) and very high (>30 ppm) available P was 10.1, 79.9, and 10%, respectively (Table 1, Fig 1). More area under higher value of available P as compared to number of samples was for similar reason as in case of organic carbon. This further indicated that in potato based system due to poor efficiency of P in potato crop and regular application to all crops of the system has resulted into it's high build up in the soil. Many earlier works has also reported positive balance of P in the potato based systems.

More than 89% of area under potato growing pockets were low in available K (<105 ppm) for potato crop as compared to 88% of soil samples (Table 1, Fig 1)

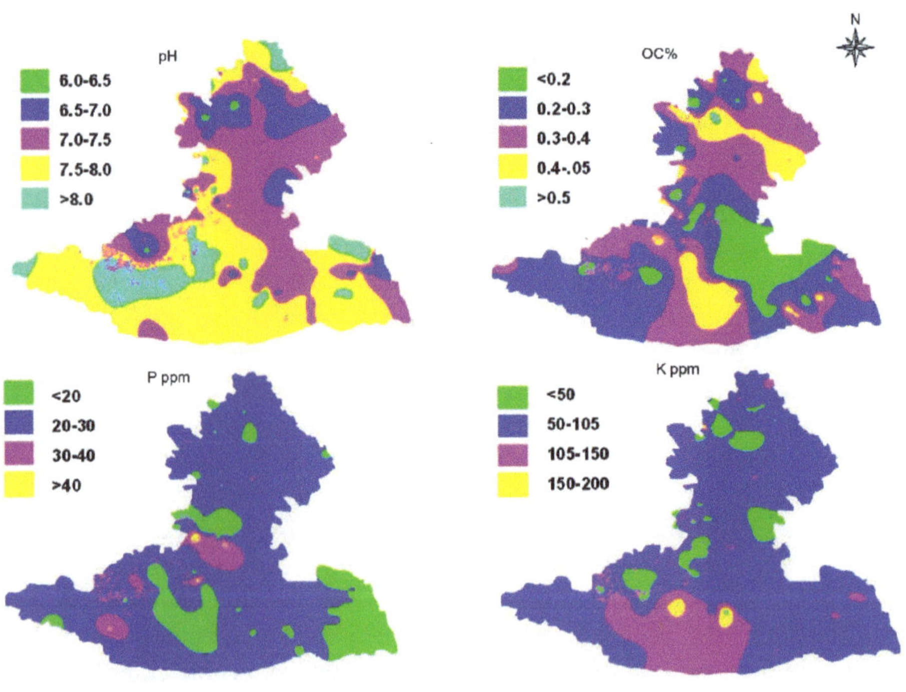

Fig. 1 : Spatial distribution of soil pH, organic carbon (OC%), available P and available K in potato growing pockets of Jalandhar district of Punjab. (pink pixels represent potato area)

which was the result of very intensive cultivation, along with sandy loam texture and poor cation holding capacity of soils in these pockets. Potato crop is one of the most exhaustive crop to soil potassium and potato based cropping system mostly have negative balance for this nutrient. In these pockets there appears very severe pressure on soil K reserve and to maintain the reserve there is need to apply as much K as to replenish the amount removed by the crops of the system.

Table 1: Per cent soil samples from potato fields and potato growing area falling in different ranges of pH, organic carbon (%), available P and K.

pH		OC (%)		Available P (ppm)		Available K (ppm)	
Range	**% samples**	**Range**	**% samples**	**Range**	**% samples**	**Range**	**% samples**
6.0-6.5	5.1	0.2	19.8	<10	0.6	<105	87.9
6.5-7.0	15.3	0.2-0.3	28.0	10-20	30.6	105-150	10.2
7.0-7.5	24.2	0.3-0.4	31.2	20-30	58.0	150-200	0.6
7.5-8.0	33.8	0.4-0.5	18.5	>30	10.8	>200	1.3
>8.0	21.7	>0.5	2.6				
Range	**% area**	**Range**	**% area**	**Range**	**% area**	**Range**	**% area**
6.0-6.5	0.7	<0.2	8.6	<20	10.1	<50	13.8
6.5-7.0	15.5	0.2-0.3	44.0	20-30	79.9	50-105	75.3
7.0-7.5	36.2	0.3-0.4	37.9	30-40	8.9	105-150	10.7
7.5-8.0	29.0	0.4-0.5	9.3	>40	1.1	150-200	0.2
>8	18.8	>0.5	0.3				

Integration of P and K map by overlaying showed that about 8.5% area was high in P and medium in K, whereas, about 81.5% area was high to very high in P but low in available K. Similarly, about 2.5% area was medium in available K but low in available P and 7.6% area was low in both available P and K. Therefore, while recommending fertilizer for these areas one must take care of soil status. Not all fields having high requirement of K need high application of P and low requirement of K need less application of P.

Micronutrients: In general percent area under micronutrients deficiency was less than the percent of samples, and percent deficient area under individual micronutrient varied from 11.1% for Mn to 14.3% for Fe (Table 2). It is important to note that not even a single field was deficient in all micronutrients; therefore, there is a need to apply these nutrients on soil test basis. Multi-micronutrients map obtained from integration of four maps of individual micronutrients made it clear that less than 1% area was deficient in all four micronutrients, about 1.78% area was deficient any of three elements (Zn, Cu, Mn; Zn, Cu, Fe; Cu, Mn, Fe or Zn, Mn Fe), 8.03% area were deficient in at least two elements and 26.69% area was deficient in only one of these four elements (Table 3). However, majority of potato area (62.92%) had sufficient levels of all four micronutrients, hence does not require immediate attention for application of micronutrients. It was clear that out of 37% area deficient in one or more nutrients majority of area (26.69%) was deficient in only one nutrient element. This becomes important management point of view as it will require selective use of different micronutrients at different location in the pockets. From these spatial distribution maps it was also evident that deficiency zones of Cu and Mn was in the area where potato fields were more concentrated, whereas, deficiency of Zn was more in the area where potato fields were not so dense. It infers that percent areas of the district deficient in Zn is likely to be much more than what indicated for potato pockets, while that may not be true for Cu and Mn.

Table 2: Per cent soil samples from potato fields and potato growing area in deficient and sufficient in available micronutrients.

Micro- nutrient	Zn	Cu	Mn	Fe	Zn	Cu	Mn	Fe
Critical limit (ppm)	0.75	0.32	3.00	6.60	0.75	0.32	3.00	6.60
	Per cent samples					Per cent area		
Deficient	29.3	15.9	13.4	23.6	13.7	11.4	11.1	14.3
Sufficient	70.7	84.1	86.6	76.4	86.3	88.6	88.9	85.7

Table 3 : Area under various multi-micronutrient deficiency categories of potato growing region in the Jalandhar district in Punjab

S. No.	Fe	Cu	Mn	Zn	Area (%)
1	D	D	D	D	0.68
2	D	D	D	S	0.90
3	D	D	S	D	0.33
4	D	D	S	S	1.79
5	D	S	D	D	0.01
6	D	S	D	S	0.40
7	D	S	S	D	4.48
8	D	S	S	S	5.68
9	S	D	D	D	0.45
10	S	D	D	S	0.27
11	S	D	S	D	0.27
12	S	D	S	S	6.72
13	S	S	D	D	0.82
14	S	S	D	S	7.61
15	S	S	S	D	6.68
16	S	S	S	S	62.92
					100.00

S= Sufficient, D= Deficient

Spatial distribution of available macro and micronutrients in the potato growing pockets of Hoshiarpur

Soil pH and organic carbon: After kriging for soil pH and classification in the six classes in the range of<6.0, 6.0 to 6.5, 6.5 to 7.0 and 7.0 to 7.5 and 7.5 to 8.0, area fell under these categories were 5.1, 27.9, 52.1, 14.1 and 0.8%, respectively (Table 4). None of the samples were found more than 8.0 pH. More than 94% of the potato soils were falling in pH range from 60 to 7.5 (neutral) a situation suitable for potato cultivation. Similarly, area under organic carbon in range 0.2 to 0.3, 0.3 to 0.4, 0.4 to 0.5 and >0.5 %, was 0.3, 19.2, 55.6 and 24.9%, respectively. None of soil sample was found in <0.2% OC categories.

Nitrogen: None of the potato growing area were found in high in available N. 63% of potato growing area was found in less than 280 kg N/ha and remaining 37% of soil samples were in more than 280 kg N/ha. Due to very intensive and exhaustive cropping systems being followed in the region which has light textured soils with poor nutrient and water holding capacity, almost no recycling of crop residue and negligible availability of FYM, it resulted into severe deterioration of soil available nitrogen.

Phosphorus: Area under medium, high and very high available P was 10.0, 39.3 and 50.6%, respectively as compared to per cent number of samples under medium (11.3%), high to very high (88.7%) (Table 4 & Fig. 2). Area under higher value of available P as compared to number of samples was same. The results indicated that in potato and potato based cropping systems due to poor efficiency of this nutrient in potato crop and regular application to all crops of the system has resulted into high build up in the soil. Due to its immobile nature in the soil it has appeared as threat to nutrient balance and also a problem for surface water runoff leading to accelerated eutrophication of water bodies.

Potassium: About 82 to 83 and 17% of area under potato growing pockets were low and medium in available K (<105 ppm and more than 105 ppm) for potato crop as compared to 79.1 and 19.21% of soil samples (Table 4 & Fig. 2). The low availability status of K is result of very intensive cultivation, along with sandy loam texture and poor cation holding capacity of soils in these pockets. Potato crop is one of the most exhaustive crop for soil potassium and potato based cropping system mostly have negative balance for this nutrient. In these pockets there appears very severe pressure on soil K reserve and to maintain that reserve there is a need to apply as much K as to replenish the amount removed by crops of the system.

Integration of N, P and K map by overlapping showed eighteen different classes of N, P and K (Table 5). The 49.24% area (30.49 and 18.75%) was under high and very high in P but low in N and K. Similarly, about 17.27% area was medium in available N and very high in P but low in available K and 12.79% area was medium in both available N and K but very high in P. Therefore, while recommending fertilizer for these areas one must take care of soil status.

The spatial classified maps generated under the study will be useful for identifying specific locale of potato growing pockets with different nutrient management problems. This study also highlighted that almost whole area is low in OC, N and K. The N and K deficiency started appearing as most of the farmers are not applying balanced fertilizers thereby suggesting a need for use of fertilizers based on soil test values. With the identification of locales with one or more macronutrients deficiency, it is possible to suggest very site specific recommendations and management. These maps can further be overlaid on the digitized boundaries of village/block/tehsil to prioritize the villages which need immediate attention of the district officials, scientists, and individual farmer in respect of application/management of required amount and kind of macro nutrients. It is also possible to monitor the changes taking place at different locations over a period of time due to different cropping systems and management practices with respect to nutrient status and other soil properties.

Table 4 : Percent of potato growing area falling in different ranges of pH, organic carbon, available K and P.

pH		OC		Available N		Available K		Available P	
Range	Area (%)	Range (%)	Area (%)	Range (%)	Area (%)	Rang (ppm)	Area (%)	Range (ppm)	Area (%)
<6.0	5.13	<0.2	0.00	<150	0.00	<105	82.33	<20	10.0
6.0-6.5	27.9	0.2-0.3	0.3	150-280	62.7	105- 150	17.1	20-30	39.3
6.5-7.0	52.1	0.3-0.4	19.2	>280	37.3	>150	0.6	>30	50.6
7.0-7.5	14.1	0.4-0.5	55.6	-	-	-	-	-	-
7.5-8.0	0.8	>0.5	24.9	-	-	-	-	-	-
>8	0.0	-	-	-	-	-	-	-	-

Fertilizer recommendations

Fertilizers application in the potato crop varies with farmers' perception about the recommendation for the area and their resources. Since farmers are not aware of the available soil nutrient soil status, application rate are not matching with the requirement of the crop. The recommendation of N for soil below 280 kg ha^{-1} available N for a yield target of 250, 300 and 350 q ha^{-1} was in the range of 70, 93 and 117 kg ha^{-1}, respectively (Table 6). However, for the same target yields when soil test values in some part of the pocket increased above 400 kg ha^{-1} available N, the recommendation decreased below 50, 73 and 97 kg ha^{-1}N, respectively. In case of P, where available P_2O_5 was between 10-20 kg ha^{-1}, recommendation of phosphorus dose for yield target of 250, 300 and 350 q ha^{-1}

ranged from 30-26, 37-33 and 44-40 P_2O kg ha^{-1}, respectively, but in the majority of area where available P was more than 50 kg ha^{-1} recommendation for P fertilizer application reduced to a meager amount. Similarly, for potassic fertilizer application, in the patches where available K was less than 300 kg/ha, fertilizer application at the rate 30, 45 and 60 kg ha^{-1} K_2O must be followed for the yield targets of 250, 350 and 350 q ha^{-1}.

Table 5 : Area under various multi micronutrient deficiency categories of potato growing region in the Jalandhar district in Punjab.

S. No.	N	P	K	Area (%)
1	L	M	L	9.38
2	L	M	M	0.21
3	L	M	H	0.00
4	L	H	L	30.49
5	L	H	M	0.32
6	L	H	H	0.00
7	L	VH	L	18.75
8	L	VH	M	3.50
9	L	VH	H	0.00
10	M	M	L	0.40
11	M	M	M	0.00
12	M	M	H	0.03
13	M	H	L	8.02
14	M	H	M	0.22
15	M	H	H	0.28
16	M	VH	L	15.27
17	M	VH	M	12.79
18	M	VH	H	0.31

L: Low, M: Medium, H: High, VH: Very high

The spatial classified maps generated under the study will be useful for identifying specific locale of potato growing pockets with different nutrient management problems. While large area came under high P, requiring low or less application of this nutrient, this study also highlighted almost whole area low in OC indicating severe N deficiency and identified locales with low available K. With the identification of locales with one or more macro/ micronutrients deficiency it is possible to suggest very site specific recommendations and management. These maps can further be overlaid on the digitized boundaries of village/block/tehsil to prioritize the villages which need immediate attention of the district officials, scientists, and individual farmer in respect of application/management of required amount and kind of macro and micronutrient elements. Further, with the use of prescription equations yield target specific recommendation may also be given for all fertilizer nutrients. It is also possible to monitor the changes taking place at different locations over a period of time due to different cropping systems and management practices with respect to nutrient status and other soil properties. In addition such studies may be extended to map spatial variability of available nutrients of large number of pockets in order to progress towards precision farming (Kumar *et al.*, 2014, Jatav *et al.*, 2013 and Kumar *et al.*, 2015).

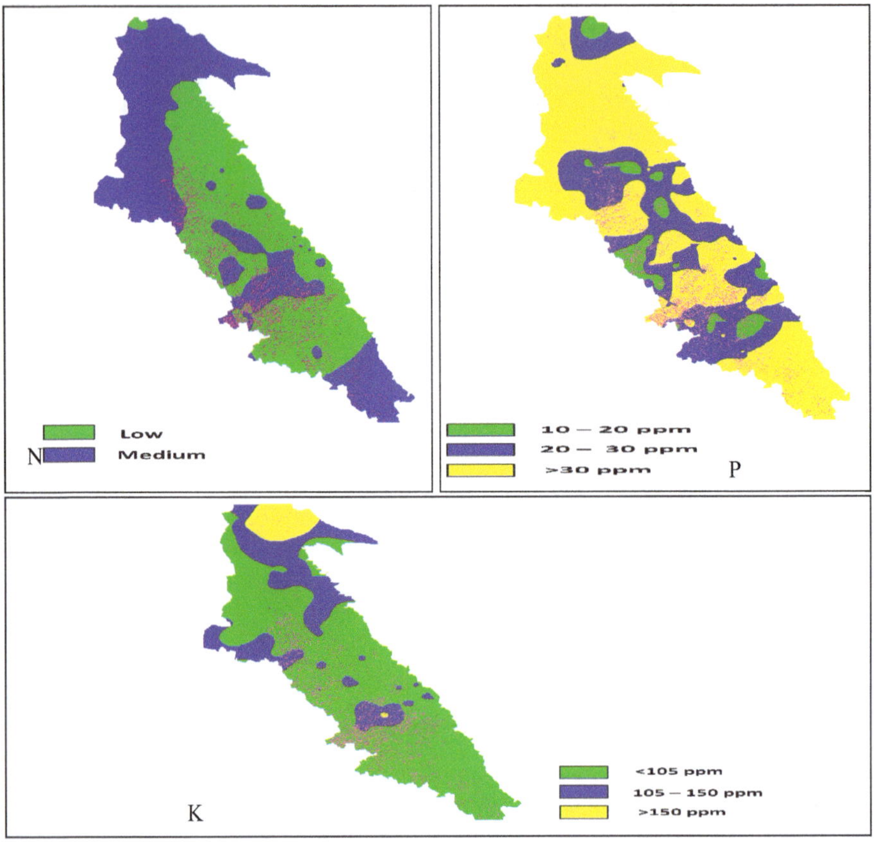

Fig. 2 : Spatial distribution of available N, available P and available K in potato growing pockets of Jalandhar district of Punjab. (pink pixels represent potato area)

Table 6 : Fertilizers recommendation based on target yield equation for potato growing areas

Soil N (kg/ha)	N commendation based on target yield equation		
	250 q/ha	300 q/ha	350 q/ha
< 100	227	287	348
100-200	227-151	287-211	348-272
200-300	151-75	211-135	272-196
300-400	75-0	135-58	196-120
400-500	-	58-0	120-44
Soil P (kg/ha)	**P_2O_5 commendation based on target yield equation**		
	250 q/ha	300 q/ha	350 q/ha
5-15	95-81	117-103	139-125
15-25	81-51	103-73	125-95
25-50	51-22	73-44	95-66
50-70	22-0	44-15	66-37
70-100	0	0	8-0
Soil K (kg/ha)	**K_2O commendation based on target yield equation**		
	250 q/ha	300 q/ha	350 q/ha
<	190	258	326
100-150	190-115	258-183	326-251
150-200	115-40	183-33	251-176
250-300	40-0	33-0	176-101
300-350	-	-	101-26

References

Agarwala S.C. and Sharma C.P. (1979). Recognizing Micronutrient Disorders of Crop Plants on the basis of Visible Symptoms and Plant Analysis. ICAR, New Delhi.

Aggarwal, R.K.; Kumar, P. and Sharma, B.K. (1990). Distribution of nitrogen in some Aridisols. *J. Indian Soc. Soil Sci.*, **38**: 430-433.

Aggarwal, R.K. and Kumar, P. (1994). Availability and management of nitrogen in soils of arid ecosystems. *Ann. Arid Z one*, **33(1)**: 1-18.

Akbari, K.N.; Fateh Karan and Pandya, H.R. (1993). Distribution of available P, K and organic carbon in Mewar region. *Ann. Arid Zone*, **32(4)** : 251-252.

Alloway B.J. (1995). Heavy Metals in Soils, 2nd edition, Wiley, New York.

Alloway B.J. (1995a). Soil processes and their behaviour of metals. In: Alloway B.J. (ed), Heavy Metals in Soils, 2nd edition. Wiley, New York, pp. 11-38.

Anonymous, (2002). *Annual Report* (*rabi*). Division of Olericulture, SKUAST (K), Shalimar Srinagar.

Anonymous, (2003). *Annual Report* (*rabi*). Division of Olericulture, SKUAST (K), Shalimar Srinagar.

Arnon D.I. and Stout P.R. (1939). The essentiality of certain elements in minute quantity for plants with special reference to copper. Plant Physiol. 14: 371-385.

AVRDC, (2004). AVRDC Medium Term Plan: 2004-2006. Shanhua, Taiwan: AVRDC- the World Vegetable Center. AVRDC Publication 04-567, pp. 46.

Ayers, R.S. and Westcot, D.W. (1976). Water quality for agriculture. Irrigation and Drainage paper, 29, FAO, Rome, 97 p.

Bagyaraj D.J. and Sreeramulu K.R. (1982). Pre inoculation with *vesicular-arbuscular* mycorrhiza improves growth and yield of chilli transplanted in the field and saves phosphatic fertilizers. *Plant and Soil* **69**: 375-382.

Bairwa H.L., Shukla A.K., Mahawer L.N., Kaushik R.A., Shukla K.B. and Ameta K.D. (2009). Response of integrated nutrient management on yield, quality and physico-chemical characteristics of okra cv. Arka Anamika. *Indian Journal of Horticulture* **66**(3): 310-314.

Baligar V.C., Fageria N.K. and Elrashidi M.A. (1998). Toxicity and nutrient constraints on root growth. *Hort Sci.* 33: 960-965.

Bansal R.L., Nayyar V.K., Takkar P.N. and Brar J.S. (1986). Status of available Zn, Cu, Fe and Mn in three soil associations of Kapurthala district (Punjab). *Journal of Research of Punjab Agricultural University* **23**: 193-200.

Barber S.A. (1995). Soil Nutrient Bioavailability: A Mechanistic Approach, 2nd edition. Wiley, New York.

Belanger R.R., Bower P.A., Ehert D.S. and Menzies J.G. (1995). Soluble silicon: Its role in crop and disease management of greenhouse crops. *Plant Dis.* 79: 329-336.

Bennett W.F. (1993). Plant nutrient utilization and diagnostic plant symptoms. In: Bennett, W.F.(ed.), Nutrient Deficiencies and Toxicities in Crop Plant. American Phytopathological Society Press, St. Paul, MN, pp. 1-7.

Bergmann W. and Neubert P. (1976). Plant Diagnosis and Plant Analysis veb sustav fischer verlag, Jena.

Bhargava, B.S. and Raghupathi, H.B. (1993). In methods of analysis of soils, plants waters and fertilizers. H.L.S. Tandon (Ed) F.D.C.O., New Delhi, pp. 41.

Bhargawa, B.S. and Chadha, K.L. (1993). Leaf nutrient guide. In *Advances in Fruit Crops*, Malhotra publishers, New Delhi, India.

Bhattarai I.D. and Mishra R.R. (1984). Study on the *vesicular-arbuscular mycorrhiza* of three cultivars of potato (*Solanumtuberosum* L.). *Plant and Soil* **79**: 299-303.

Bould C., Hewitt E.J. and Needham P. (1983). "Diagnosis of mineral disorders in plants, Vol. I, Principles" Her Majesty's Stationary Office, London.

Brady and Weil, (2002). *Elements of the Nature and Properties of Soil. Prentice Hall, New Jersey.*

Bray, R.H. and Kurtz, L.T. (1945). Determination of total, organic and available forms of phosphorus in soils. *Soil Sci.* **59:** 39-45.

Brookes, P.C., Landman, A., Pruden, G. and Jenkinson, D.S. (1985). Chloroform Fumigation and the release of soil nitrogen: a rapid direct extraction method to measure microbial biomass nitrogen in soil. *Soil. Boil. Biochem*. **17**: 837-842.

Brown J.C., Ambler J.E., Chaney R.L. and Foy C.D. (1972). Differential responses of plant genotypes to micronutrients. In: Mortvedt JJ, Cox FR, Shuman LM and Welch RM (ed), Micronutrients in Agriculture, 2nd edition, SSSA, Madison, WI, pp. 389-418.

Brown P.H. and Shelp B.J. (1997). Boron mobility in plants. *Plant Soil,* 193: 85-101.

Casida, I.E., Klein, D.A. and Santore, T. (1964). Measurement of dehydrogenase activity by incubating the soil with TTC method. *Soil Sci.*, 98: 373.

Cate, R.B. Jr. and Nelson, L.A. (1965). A rapid method for correlation of soil test analyses with plant response data. Tech. Bull. 1. N. Carol, State Agric. Exp. Stn. ISTP Series.

Chatto M.A., Gandorio M.Y. and Zargar M.Y. (1997). Effect of *Azospirillum* and *Azotobacter* on growth and quality of knolkhol (*Brassica oleracea* L. Var. gongylodes). *Vegetable Science* **24**: 16-19.

Chattopadhyay, T., Singh, R.S., Sahoo, A.K. and Shyampul, R.L. (1997). Available micronutrient status of Rajasthan soils. *Agropedology*, 7: 40-43.

Chesnin, L. and Yien, C.H. (1957). *Proc. Soil Sci. Amer.,* 15: 149.

Choudhury M.L., Rajput C.B.S. and Ram H. (1982). Effect of *Azotobacter* and *Rhizobium* treatments on growth, yield, and quality of garden pea. *Haryana Journal of Horticultural Science* **11**: 231-234.

Clark R.B. and Zeto S.K. (2000). Mineral acquisition by Arbuscular mycorrhizal plants. *J. Plant Nutri.* 23: 867-902.

Cox F.R. (1987). Micronutrient soil tests: correlation and calibration. In: Brown JR (ed), Soil Testing: Sampling, Correlation, Calibration and Interpretation. Soil Sci. Soc. Am., Madison, WI, pp. 97-117.

De Nirmal, Singh K.P. and Rai M. (2004). Micronutrient deficiency symptoms in vegetable crops and their management. Tech. Bull. No. 15, IIVR, Varanasi. pp. 24.

De Nirmal, Singh K.P. and Rai M. (2004a). Micronutrients in Vegetable. Tech. Bull. No. 22, IIVR, Varanasi, pp. 10.

De Nirmal and Rai Mathura. 2005. Micronutrient disorders and management in Uegetable crops: A Review. Veg. Sci. 32 (1): 1-10.

Delgado and Lemunyon, (2006). Nutrient Management. In *Encyclopedia of Soil Science* Rattan Lal (Ed.). pp. 1157-1160.

Desmond G.M. and Walter A.H. (1990). Sweet potato growth and nitrogen content following nitrogen application and inoculation with *Azospirillum. Horticulture Science* **25**: 758-759.

Dhir, R.P., Sharma, B.K. and Joshi, D.C. (1983). Availability of iron, manganese, zinc and copper in some soils of western Rajasthan. *Ann. Arid Zone,* **22(4)**: 343-349.

Dhir, R.P., Singh, N. and Sharma, B.K. (1979). Nature and indices of soil salinity in Pali block, western Rajasthan. *Ann. Arid Zone,* **18:** 27-34.

Dogne, J.D. (1995). Studies on the physico-chemical properties of problematic soils of Kanwar Sen Lift Canal area of IGNP. M.Sc. (Ag.) Thesis, R.A.U., Bikaner.

Doneen, L.D. and Headerson, U.W. (1960). Quality of irrigation water and chemical and physical properties of soils. *Trans. 7th Int. Congs. Soil Sci.,* 1: 516-522.

Duncan R.R. (1994). Genetic manipulation. In: Wilkinson RE (ed), Plant Environmental Interactions, Dekker, NY, pp. 1-38.

Duncan R.R. and Carrow R.N. (1999). Turf grass molecular genetics improvement for abiotic/edaphic stress resistance. Adv. Agron. 67: 233-306.

Dutta, B.K. and Joshi, D.C. (1989). Status and availability of potassium in sandy soils of arid Rajasthan. International Symp. of ISSS on "managing sandy soils" (Abst.) held from Feb. 6-10 at CAZRI, Jodhpur. pp. 265.

Engelhard A.W. (1990). Soil borne Plant Pathogen, Management of Disease with Macro and Micro Elements. American Phytopathological Society Press, St. Paul, MN.

Epstein E. (1994). Review: The anomaly of silicon in plant biology. Proc. Natl. Acad. Sci. USA. 91:11-17. Retrieved Nov. 11, 2012 from http://www.pnas.org/content/91/1/11.full.pdf.

Erisman, J.W., Sutton, M.A., Galloway, J.N., Klimont, Z. and Winiwarter, W. (2008). How a century of ammonia synthesis changed the world. *Nat. Geosci.* 1: 636-639.

Ess D.R., Morgan M.T., Parsous S.D. (2003). Implementing site-specific management: map-versus sensor-based variable rate application. http://www-agcom,purdue.edu/Ag.Com/Pubs/SSM-2-W. pp 1-9.

Fageria N.K., Baligar V.C. and Clark R.B. (2002). Micronutrients in crop production. *Adv. Agron.* 77: 185-268.

Fageria N.K., Baligar V.C. and Jones C.A. (1997). Growth and mineral nutrition of field crops. 2nd Ed. Dekker.

Fageria, N.K. and Zimmermann. F.J.P. (1998). Influence of pH on growth and nutrient uptake by crop species in an Oxisol. *Communications in Soil Science and Plant Analysis* 29 (17&18): 2675–2682.

Fairweather J.V. and D.J. Parbery, (1982). Effects of four *vesicular-arbuscular mycorrhizal* fungi on growth of tomato. *Transactions of the British Mycological Society* **79**: 151-153.

Ganeshamurthy, A.N., Kalaivanan, D. and Selvakumar, G. 2015. Nutrient management in horticultural crops. India J. Fert., Vol: 11(12), pp. 30-42

Ganeshamurthy, A.N., Satisha, G.C. and Prakash Patil, (2011). Potassium nutrition on yield and quality of fruit crops with special emphasis on banana and grapes. *Karnataka J. Agric. Sci.,* 24(1): 29-38.

Gerloff G.C. and Gabelman W.H. (1983). Genetic. Genetic basis of inorganic plant nutrition. In: Lauchli A and Bieleski RL (ed), Inorganic Plant Nutrition. Encyclopedia of plant physiology, New Series, Vol. 15B Springer-Verlag, New York. pp. 453-480.

Gomaa A.M.H. (1995). Response of certain vegetable crops to bio-fertilization. *Ph. D. Thesis*, Faculty of Agriculture, Cairo University, Egypt.

Goswami N.N. (2006). Soil testing as a unique tool for integrated and balanced fertilizer use for optimizing production – issues, options and limitations. Key-note Address. Proc. Of the National Seminar on Soil Testing for Balanced and Integrated Use of Fertilizers, New Delhi, p. 3-19.

Goswami N.N., Subba Rao A., Biswas A.K. (2009). Soil testing for precision agriculture. Soil Science in Meeting the Challenges to Food Security and Environmental Quality. pp. 47-54.

Graham R.B. and Webb M.J. (1991). Micronutrients and disease resistance and tolerance in plants. In: Mortvedt JJ, Cox FR, Shuman LM and Welch RM (ed) Micronutrients in Agriculture, 2nd edition, SSSA, Madison, WI. pp. 329-370.

Graham R.D. (1984). Breeding for nutrition characteristics in cereals. *Adv. Plant Nutri.* 1: 57-102.

Graham R.D., Ascher J.S. and Hznes S.C. (1992). Selecting zinc efficient cereal genotypes for soils of low zinc status. *Plant Soil,* 146: 241-250.

Grant, R.F. (1991). The distribution of water and nitrogen in the soil-crop-system: A simulation study with validation from a winter wheat field trial. *Fert. Res.* **27**: 199-213.

Grigg, G.I. (1953). A rapid method for the determination of molybdenum in soils. *N.Z. Sci. Technol.,* 34: 405.

Groot, J. J. R. (1987). Simulation of nitrogen balance in a system of winter wheat and soil. Simulation Report 13, Centre for Agrobiological Res./Dept. of Theoretical production. Ecol., Agril. Univ, Wageningen.

Gupta U.C. (1979). Some factors affecting the determination of hot-water-soluble boron from podzol soils using azomethine-H. *Can. J. Soil Sci.* 59: 241–247.

Gupta, U.C. (1993). Factors affecting boron uptake by plants. In: Gupta UC (ed) Boron and its role in crop production, CRC Press, Boca Raton, FL. pp. 87-123.

Gurubatham J.P., Thamburaj S. and Kandaswamy S. (1989). Studies on the effect of bio-fertilizers on the bulb yield in Bellary Onion (*Allium cepa*). *South Indian Horticulture,* **37**: 150-153.

Hand Book No. 60, Oxford and IBH Publishing Co., New Delhi.

Hatcher, J.T. and Wicox, L.V. (1950). Colorimetric determination of boron using carmine. *Annal.Chem.* 22: 567.

Havlin, J.L., Beaton, J.D., Tisdale, S.L. and Nelson, W.L. (2005). Soil *Fertility and Fertilizers: An Introduction to Nutrient Management.* Upper Saddle River, NJ: Pearson Education (2005).

Hopkins W.G. (1999). Introduction to Plant Physiology. 2nd ed. New York, NY: John Wiley & Sons, Inc. p. 61-76.

Horesh I. and Levy Y. (1981). Response of iron-deficient citrus trees to foliar iron sprays with a low-surface-tension surfactant. *Sci. Hort.* 15: 227-233.

Huber D.M. (1980). The role of mineral nutrition in defense. In: Hosfall JG and Cowling EB (ed), Plant Pathology: An Advanced Treatise, AP, NY. pp. 381-406.

Huber D.M. and Wilhelm N.S. (1988). The role of manganese in resistance to plant diseases. In: Graham RD, Hannam R and Uren NC (ed), Manganese in Soils and Plant Kluwer Academic, Dordrecht, The Netherlands. pp. 155-173.

Hussey R.B., Peterson R.L. and Tiesson H. (1984). Interaction between *vesicular-arbuscular mycorrhizal* fungi and asparagus. *Plant and Soil* **79**: 403-415.

Jackson M.L. (1967). Soil Chemical Analysis. Prentice Hall of India Pvt. Ltd., New Delhi.

Jackson, M.L. (1973). Soil chemical analysis. Prentice Hall of India Pvt. Ltd., New Delhi.

Jatav, M.K. Dua, V.K. Kumar, Manoj, Trehan, S.P. and Kumar, Sushil, (2013). Spatial distribution of soil available nutrients in the potato growing pockets of Hoshiarpur district of Punjab. *Potato J.* 40(2): 128-134

Jatav, M.K., Dua, V.K. Kumar, Manoj, Khurana, S.C., Bhatia, A.K. Nandekar, D.N., Manorama, K., Trivedi, S.K., Das, S.N., Chettri, M, Raghav, Manoj, Verma, R.B., Naik, P.S. and Verma, Dharminder, (2012). Contribution of nitrogen and phosphorus from combined application of *Azotobacter* and phosphobacteria in potato *Vegetable Science,* 39(2): 123-127.

Jatav, M.K., Dua, V.K., Kumar, Manoj, Trehan, S.P., Singh, S.K., Khurana, S.C., Bhatia, A.K., Nandekar, D.N., Manorama, K., Patel, C.K., Trivedi, S.K., Das, S.N., Chettri, M., Raghav, Manoj, Jha, Gokelesh, Saikia, Mrinal, Naik, P.S. and Verma. Dharminder, (2013). Soil test based phosphorus recommendation for potato. *Vegetable Science*, 40(1): 17-22.

Jatav, M.K., Kumar, Manoj, Dua, V.K. and Lal, S.S. (2011). Bio-fertilizers for sustaining potato productivity in rainfed hills. *Journal of Eco-friendly Agriculture,* 6(1): 1-5.

Jatav, M.K., Kumar, S., Dua V.K, Kumar M. and Trehan, S.P. (2016). Quantification of contribution of PSB in reducing the requirement of P fertilizer in potato production in mid hill of Himachal Pradesh. *J. of Agril. & Ecology*, 1: 81-84.

Jatav, M.K., Sharma, B.D., Samadia, D.K. and Meena, S.R. (2015). Effect of micronutrient application on Mateera (*Citrullus lanatus)* Under Hot Arid Agro-climate. *Annal of Arid Zone* June, 54(3&4): 133-135.

Jatav, M.K., Sharma, B.D., Samadia, D.K. and Meena, S.R. (2016). Site specific nutrient management of kachri using omission plot technique in hot arid region of Rajasthan. *International Journal of Basic and Applied Biology*, 3: 173-175.

Jatav, M.K., Sharma, B.D., Samadia, D.K. and Meena, S.R. (2016). Efficacy of *Rhizobium* biofertilizer inoculation on graded N levels and net return from cluster bean production under hot arid region. *Economic Affairs:* EA: 61(3): 495-499.

Jatav, M.K., Sharma, B.D., Samadia, D.K. and Meena, S.R. (2016). Effect of different sources of nitrogen on *Kachari* (*Cucumis melo*) performance in the arid region. *Annals of Arid Zone* 55(1&2): 35-39.

Jatav, M.K., Singh B.P., Kumar, Manoj, Dua,V.K., Singh, S.K., Khurana, S.C., Bhatia, A.K., Singh, Name, Nandekar, D.N., Manorama, K., Patel, C.K., Siddagangaiah G, Trivedi, S.K., Das, S.N., Chettri, M., Jha, Gokelesh, Naik, P.S., Verma, D., Bairwa, Ishwar Babu, Kumar, Pankaj, Kumar Lalit and Choudhary, Anirudh, (2015). Performance of Phosphobacteria on Phosphorus Economy and Net Return in Different Potato Growing Region of India. *Economic Affairs*, 60(4): 727-734.

Jatav, M.K.; Dua, V.K.; Kumar, Manoj, Kumar, Sushil, Kumar, Bablesh and Kumar, Lalit, (2015). Contribution of azotobacter in reducing the requirement of nitrogen in potato production in mid hill of Himachal Pradesh. *Progressive Research – An International Journal*, 10 (Special-V): 2874-2877.

Kameriya, P.R. (1995). Characterization of soils of Agro-climatic zone of transitional plain of inland drainage (Zone IIa) of Rajasthan. Ph.D. Thesis, R.A.U., Bikaner.

Kanaujia S.P., Tripathy D., Narayan R. and Shukla Y.R. (1999). Influence of phosphorus, potassium, and *Rhizobium* on green pod yield of pea. *Advances in Horticulture and Forestry,* **7**: 107-112.

Kannan S. (1990). Role of foliar fertilization on plant nutrition. In: Baligar VC and Dunean RR (ed), Crops as enhancer of nutrient use. AP, San Diago, pp. 313-348.

Kanwar J.S. (1976). Micronutrients. In: Kanwar JS (ed), Soil Fertility – Theory and Practice. ICAR, New Delhi, pp. 229-300.

Kanwar J.S. and Randhawa N.S. (1967). Micronutrient Research in Soils and Plants. (A Review), ICAR, New Delhi.

Karamanos R.E., Kruger G.A. and Steward J.W.B. (1986). Copper deficiency in cereal and oilseed crops in northern Canadian praine soils. *Agron. J.* 78: 317-323.

Karuthamani M., Natrajan S. and Thamburaj S. (1995). Effect of inorganic and bio-fertilizers on growth, flowering and yield of pumpkin. *South Indian Horticulture,* **38**: 345-346.

Kenworthy, A.L.(1964). *Fruit, Nut and Plantation Crops. A guide for collection of foliar samples for nutrient element analysis*. Hort. Dept. Memo. Michigan State Univ., East Lansing.

Khan S. and Pariari A. (2012). Effect of N-fixing Bio-fertilizers on growth, yield and quality of chilli (-*Capsicium annum* L.). *The Bioscan,* **7**(3): 481-482.

Kishor, R. (2003). Salinity and alkalinity indices of soils irrigated with saline water in Rajgarh tehsil of Churu district (Rajasthan). M.Sc. (Ag.) Thesis, R.A.U., Bikaner.

Kobata – Pendias A. and Pendias H. (1992). Trace Elements in Soils and Plants. 2nd edition. CRC Press, Boca Raton, FL.

Kotur, S.C. (1990). Mo nutrition of cauliflower as affected by Mo application and liming on an acid sandy loam soil. *J. Indian Society of Soil Sci*., 38(2): 280.

Kovda, V.A. Yaron, B. and Skalhevat, J. (1967). Quality of irrigation water. An international source book on irrigation and drainage of arid lands in relation to salinity and alkalinity. FAO/ UNESCO pp. 246-282.

Kucey R.M.N., Janzen H.H., Leggett M.E., (1989) Microbially mediated increases in plant available phosphorus. *Adv. Agron.* **42**: 199–225.

Kumar, Mahesh; Panwar, N.R.; Pandey, C.B. and Jatav, M.K. (2015). Spatial distribution of available nutrients in the potato growing areas of Banaskantha district of Gujarat. *Potato J.* 42(2): 130-136.

Kumar, Mahesh, Jatav, M.K., and Kumar Pankaj. 2016. Soil nutrient mapping for site specific fertilizer recommendations. Good management practices for horticulture crops edited by Jatav et al 2016. NIPA New Delhi India pp.119-125.

Kumar, Manoj, Jatav, M.K., Dua, V.K., Trehan, S.P., Lal, S.S. and Upadhyay. N.C. (2014). Methodology to map spatial variability of available nutrients in area of intensively growing potato (*Solanum tuberosum*) using remote sensing and GIS. *Indian Journal of Agricultural Sciences* 84(3): 396–400, March 2014/Article.

Kumar, V. and Seth, S.P. (1983). Studies on soil fertility status of Sriganganagar district. *Ann. Arid Zone*, **22(1)** : 77-81.

Kumar, V., Mathur, G.S. and Seth, S.P. (1973). Preliminary studies on the fertility status of the soils of Sriganganagar district of Rajasthan. *Ann. Arid Zone*, **12(1&2)**: 59-64.

Kumar, V., Mohd. Fakhruddin and Seth, S.P. (1974). Changes in soil fertility in soil of Sriganganagar district of Rajasthan. *Ann. Arid Zone*, **13(1)**: 17-22.

Kumaran S.S., Natarajan S. and Thamburaj S. (1998). Effect of organic and inorganic fertilizer on growth, yield and quality of tomato, *South Indian Horticulture* **46**: 203-205.

Lake D.L., Kisk P.W.W. and Lester J.N. (1984). Fractionation, characterization and speciation of heavy metals in sewage sludge and sludge-amended soils: A reviews. *J. Environ. Qual.* 13: 175-183.

Lal, P., Singh, M. and Singh, K.S. (1971). Fertility, salinity and alkalinity indices of the soils of Jobner tract. *Rajasthan Agriculturist*. **X –XI**: 6-9.

Lindsay W.L. (1991). Inorganic equilibria affecting micronutrients in soil. In: Mortvedt JJ, Cox FR, Shuman LM and Welch RM (ed), Micronutrients in Agriculture, 2nd edition. SSSA, Madison, WI, pp. 89-112.

Lindsay W.L. and Norwell W.A. (1978). Development of DTPA soil test for Zn, Fe, Mn and Cu. *Soil Science Society of America Proceeding,* **33**: 62–8.

Lindsay, W.L. and Norvell, W.A. (1978). Development of DTPA soil test for zinc, iron, manganese and copper. *Soil Sci. Soc. Amer. J.,* 42: 421-428.

Lundegardh tr. Mitchell (1951). (London: Hilger and Watts, Ltd.) (cited in Jackson ML. 1967. Soil Chemical Analysis. Prentice Hall of India Pvt. Ltd. New Delhi).

Mallick MFR and Muthukrishnan C.R. (1980a). Effect of micronutrients on tomato (Lycopersicon esculentum Mill). II. Effect on flowering, fruit-set and yield. *South Indian Horticulture*. 28(1): 14-20.

Marschner H. (1995). Mineral nutrition of higher plants. Academic Press. San Diego.

Martens D.C. and Lindsay W.L. (1990). Testing soils for copper, iron, manganese and zinc. In: Westerman RL (ed), Soil Testing and Plant Analysis. SSSA, Madison, WI, pp. 229-273.

Martens D.C. and Westermann D.T. (1991). Fertilizer applications for correcting micronutrient deficiency. In: Mortvedt JJ, Cox FR, Shuman LM and Welch RM (ed), Micronutrients in Agriculture, 2nd edition. SSSA, Madison, WI, pp. 549-592.

Mathur, G.S. (1986). Study on characterization and effect of some reclamative measures on salt affected soils of Indra Gandhi Nahar Area. Ph.D. Thesis. Univ. of Rajasthan, Jaipur.

McBride M.B. (1994). Environmental Chemistry of Soils. Oxford University Press, New York.

Mengel K. and Kirkby E.A. (1982). Principles of Plant Nutrition, 3rd edition. International Potash Institute, Berm, Switzerland.

Menzies J.G. and Belanger R.R. (1996). Recent advances in cultural management of diseases of greenhouse crops. *Can. J. Plant Pathol.* 18: 186-193.

Metson, A.J. (1956). Methods of chemical analysis for soil survey samples. Bull. N.Z. Deptt. Scient. *Md. Res. Soil Bur.,* 12.

Mishra B.K., Sharma R.K. and Nagarajan S. (2004). Plant breeding: A component of public health strategy. *Current Sci.* 86(9): 1210-1215.

Moghe, VB., Talati, N.R. and Mathur, C.M. (1964). A modified EDTA-method for determination of soluble sulphates in soils and waters. *Curr. Sci.,* 23(8): 242.

Mortvedt J.J. (1994). Needs for controlled-availability micronutrient fertilizers. Pest Res. 38: 213-221.

Mortvedt J.J. (2000). Bioavailability of micronutrients. In: Sumner ME (ed), Hank book of Soil Science. pp. D71D88.

Mortvedt J.J., Cox F.R., Shuman L.M. and Welch R.M. (1991). Micronutrients in Agriculture, 2nd edition. SSSA, Madison, WI.

Muhr, G.R., Datta, N.P., Shankarasubramoney, H., Leley, V.K. and Donahue, R.L. (1965). Soil Testing in India. U.S.D.A. Publication, pp. 120.

Murumkar D R and Patil P L. 1996. VAM-diazotrophs bell pepper symbiosis. *Journal of Maharashtra Agricultural Universities* **21**: 394-397.

Nambiar, K.K.M. (1994). Soil Fertility and Crop Productivity under long term fertilizer use in India., ICAR, New Delhi.

Olesen, S.R., Col, S.C.W., Wantable, P.S. and Dean, L.A. (1954). Estimation of available phosphorus in soils by extraction with sodium bicarbonate USDA. *Circ.* 939: 18.

Olsen S.R. (1972). Micronutrient application. In: Mortvedt JJ, Cox FR, Shuman LM and Welch RM (ed), Micronutrients in Agriculture, 2nd edition. SSSA, Madison, WI, pp 243-264.

Olsen, S.R., Cole, C.V., Watanabe, F.S. and Dean, L.A. (1956). Estimation of available phosphorus in soils by extraction with sodium bicarbonate, U.S.D.A. *Circ.* 939.

Osakia, M., Watanabe, T. and Tadano. T. (1997). Beneficial effect of aluminum on growth of plants adapted to low pH soils. *Soil Science and Plant Nutrition* 43(3): 551-563.

Parker, F.W., Nelson, W.L., Winter, E. and Miles, I.E. (1951). The broad interpretation of soil test information. *Agron. J.* **43:** 105-12.

Penning De Vries, F.W.T., Jansen, D.M., ten Berge, H.F.M. and Bakerna, A. (1989). Simulation of eco physiological process of growth in several annual crops. Simulation monographs29, IRRI/PUDOC Publication.

Piper, C.S. (1950). Soil and plant analysis, The University of Adelaide, Australia.

Prakash, C. (2001). Fertility, salinity and sodicity indices of soils of Kanwar Sen Lift Canal area of IGNP. Ph.D. Thesis, R.A.U., Bikaner.

Raghupathi, H.B., Ganeshamurthy, A.N. and Ravishankar, H. (2014). Comparison of DRIS ratio norms of selected fruit crops. *Indian J. Hort.* 71(2): 168-175.

Rahul, D.S. and Ram Deo. (1979). Distribution of inorganic phosphorus fraction in Rajasthan soils. *Ann. Arid Zone,* **18(1&2)**: 56-57.

Ram, D. (1998). Characterization of soils of agro-climatic zone of flood prone eastern plain (IIIb) of Rajasthan. Ph.D. Thesis, R.A.U., Bikaner.

Ram, M. (1996). Fertility, salinity and alkalinity indices of soils and quality of irrigation water of Pali tehsil of Pali district (Rajasthan). M.Sc. (Ag.) Thesis, R.A.U., Bikaner.

Ramanmoorthy, B. and Bajaj, J.C. (1969). Available N, P and K status of Indian Soils. *Fert. News,* **14(8):** 24-26.

Ramanmoorthy, B. Narshimham, R.L. and Dinesh, R.S. (1967). Fertilizers application for specific yield target of Sonara-64. *Indian Farming,* **17**(5): 43-45.

Rao, K.V.G.K. and Kamra, S.K. (1991). Report of the visit to IGNP command area. 26.8.1991 to 31.08.1991.

Rao, V.J.M., Deshpande R. (1971). Effect of micronutrients (copper and boron) on the growth and yield of onion (*Allium cepa*). *Ind J. Agric. Res.* 5(4): 257-260.

Rathore, Mala (2009). Nutrient content of important fruit trees from arid zone of Rajasthan *Journal of Horticulture and Forestry*, 1(7): 103-108.

Reuter, D.J. and Robinson, J.B. (1986). *Plant Analysis and Interpretation Manual.* Inkata Press, Melbourne, Sydney.

Richards, L.A. (1954). Diagnosis and improvement of saline and alkali soils, USDA Hand Book No. **60**. Washington, D.C.

Robert L. Mikkelsen, (2011). The "4R" Nutrient Stewardship Framework for Horticulture. *Horttechnology*, 21(6): 658-662.

Romheld V. (1991). The role of phytosiderophores in acquisition of iron and other micronutrients in graminaceous species : An ecological approach. Plant Soil. 130: 127-134.

Romheld V. and Marschner H. (1986). Mobilization of iron in the rhizosphere of different plant species. *Adv. Plant Nutr.* 2: 155-204.

Romheld V. and Marschner H. (1991). Functions of micronutrients in plants. In: Mortvedt JJ, Cox FR, Shuman LM and Welch RM (ed), Micronutrients in Agriculture, 2nd edition. SSSA, Madison, WI, pp. 297-328.

Ronen, E. (2007). Micro-elements in Agriculture. *Practical Hydroponics & Greenhouses* July/August: 39-48.

Sanghi, C.L.; Lodha, B.K. and Jain, S.V. (1976). Morphological and soil fertility of western Rajasthan (Jodhpur division). *Ann. Arid Zone*, **15 (1&2)**: 23-28.

Savant N.K., Korndorfer G.H., Snyder G.H. and Datnoff L.E. (1999). Silicon nutrition and sugarcane production: *A review. Can. J. Plant Nutri.* 22: 1853-1903.

Savant N.K., Snyder G.H. and Datnoff L.E. (1997). Silicon management and sustainable rice production. *Adv. Agron.* 56: 151-199.

Scott G.J. and Suarez V. (2011). Growth rates for potato in India and their implications for industry. *Potato Journal* 36(3-4): 121-35.

Scott G.J. and Suarez V. (2012). The rise of asia as the centre of global potato production and some implications for industry. *Potato Journal* 39(3-4): 1-22.

Sehgal, J.L.; Saxena, R.K. and Vadivelu, S. (1987). Soil mapping of different states of India. *Tech. Bull.* **13,** *N.B.S.S. & L.U.P.* Nagpur, Maharastra.

Sen P., Majumdar K. and Gavin Sulewski, (2008). Importance of spatial nutrient variability mapping to facilitate SSNM in small land holding systems. *Indian Journal of Fertilizers* **4**: 43–50.

Sharma, B.D, Jatav. M.K., Balai, R.C and meens Anita. 2019. Integrated Nutrient management for horticultural crops. Dryland Horticulture Edited by Jatav et al. 2019. NIPA New Delhi. pp. 29-61.

Sharma P.K. (2004). Emerging technologies of remote sensing and GIS for the development of spatial data infrastructure. *Journal of the Indian Society of Soil Science* **52**: 384-406.

Sharma P.K., Sood Anil, Setia R.K., Verma V.K., Mehra D., Tur N.S. and Nayyar V.K. (2008). Mapping of macronutrients in soils of Amritsar district (Punjab)- A GIS approach. *Journal of the Indian Society of Soil Science* **56**: 34-41.

Sharma, G., Kuri, R.N., Dogne, R.S., Singh, Pratap, Rao. K.V.G.K. and Kamra, S.K. (1991). Report on trial and demonstration of horizontal subsurface pipe drainage. Project officer, CAD, Ground water Department, Bikaner.

Sharma, K.K. (1993). Characterization of soils of Agro-climatic zone of semi-arid eastern plain (IIIa) of Rajasthan. M.Sc. (Ag.) Thesis, R.A.U., Bikaner.

Shyampura, R.L., Singh, S.K., Singh, R.S., Jain, B.L. and Gajbhiye, K.S. (2002). *Soil Series of Rajasthan.* Technical Bulletin, NBSS&LUP, Udaipur.

Sikarwar, D.V. (1997). Fertility, salinity and alkalinity indices of soils and quality of irrigation water of Bharatpur tehsil of district Bharatpur (Rajasthan). M.Sc.(Ag.) Thesis, R.A.U., Bikaner.

Sims J.T. (2000). Soil fertility evaluation. In: Sumner ME (ed), Handbook of Soil Science. CRC Press, Boca Raton, FL, pp. D 113-D 153.

Sims J.T. and Johnson G.V. (1991). Micronutrient soil test. In: Mortvedt JJ, Cox FR, Shuman LM and Welch RM (ed),Micronutrients in Agriculture, 2nd edition. SSSA, Madison, WI, pp. 427-476.

Sindhu S.S., Tiwari R.S. (1993). Effect of micronutrients on yield and quality of onion (Allium cepa L.) cv. Pusa Red. *Progressive Hortic.* 25(3-4): 176-180.

Sindhu S.S., Tiwari R.S., Sindhu S.S. (1989). Effect of micronutrients on the growth characters of onion (*Allium cepa* L.) cv. Pusa Red: a note. *Haryana J. Hortic Sci.* 18(1-2): 146-149.

Dhyan Singh, Chhankar, P.K., and Dwivedi, B.S. (2005) Mannual on soil, plant and water analysis, 200p. Westvile, publishing House New Delhi, pp. 1-173

Singh J.P., Jaiswal V.P., Trehan S.P. (1994). In *Potato: Present and Future*. GS Shekhawat *et al.* Eds. Indian Potato Association, Shimla, 110-12.

Singh M.V. (2006). Micro- and secondary nutrients and pollutant elements research in India. Coordinator Report-AICRP Micro and secondary nutrients and pollutant elements in soil and plants, IISS, Bhopal. 30: 1-110.

Singh S. and Kapoor K.K. (1994). Solubilization of insoluble phosphates by bacteria isolated from different sources. *Environ. Ecol.* **12**: 51-55.

Singh, B. (2005). Appraisal of fertility, salinity and alkalinity indices of soils in relation to underground water of Degana tehsil of Nagaur district (Rajasthan). M.Sc. (Ag.) Thesis, R.A.U., Bikaner.

Singh, D., Rana, D.S. and Kumar, K. (1998). Phosphorus removal and available P balance in a typic Ustochrept under intensive cropping and long term fertilizer use. *Indian Journal Soc. Soil Sci.* **46(3):** 398-401.

Singh, M.V. (2008). Micronutrient Deficiencies in Crops and Soils in India. *Micronutrient Deficiencies in Global Crop Production*, 93-125. doi:10.1007/978-1-4020-6860-7_4, http://dx.doi.org/10.1007/978-1-4020-6860-7_4

Singh, S.K. and Agarwal, H.P. (1995). Vertical distribution of potassium in some normal (Ustocherepts) and salt affected (Halaquepts) soils of Jaunpur district, U.P. *J. Potash Res.* **11(1)**: 8-14.

Singh, U. (1992). Nitrogen management strategies for lowland rice cropping system, p. 110-130. In. Proc. *Int. Conf. on fertilizer usage in the Tropics*, Kuala lumpur.

Snell, F.D. and Snell, C.T. (1939). Colorimetric methods of analysis. Third Edition Vol. II D Van Nostr and Inc., New York.

Solberg E., Penny D. and Evans L. (1993). Copper deficiency in cereal crops. Agdex. 532-2.

Sood A., Sharma P.K., Tur N.S. and Nayyar V.K. (2009). Micronutrient status and their spatial variability in soils of Muktsar district of Punjab – A GIS approach. *Journal of the Indian Society of Soil Science* **57**(3): 300–6.

Splittstoesser, W.E. (1990). *Vegetable Growing Handbook: Organic and Traditional Methods* (3rd ed.). Chapman & Hall, New York.

Stevenson F.J. (1986). Cycles of soil carbon, nitrogen, phosphorus, sulfur and micronutrients.

Subbaih, B.V. and Asija, G.L. (1956). A rapid procedure for the estimation of available nitrogen in soils. *Curr. Sci.,* 25 : 259-60.

Sud K.C., Upadhayay N.C., Trehan S.P., Jatav M.K., Manoj Kumar and Lal S.S. (2008). *Soil and Plant Tests for Judicious Use of Fertilizer in Potato*. Technical Bulletin No. 91, Central Potato Research Institute, Shimla.

Sud, K.C. and Jatav, M.K. (2007). Response of potato to phosphorus and phosphorus solubilizing bacteria in brown hill soils of Shimla. *Potato J.* 34: 109-110.

Sukhada M. (1988). Nitrogen fixation in tomato *Lycopersicon esculentum* Mill. *Plant and Soil* **107**: 219-225.

Sundara B., Natarajan V. (1997). Effect of source and time of phosphorus application with and without phosphorus solubilizing bacteria on sugarcane. In: Proceedings of the Annual Convention of Sugar Technology Association of India, Goa, September 26–28, 1997, pp. 13–20.

Sundaravelu S. and Muthukrishnan T. (1993). Effect of seed treatment with Azospirillum and GA on the growth and yield of Radish. *South Indian Horticulture,* **42**: 212-213.

Swarup, A. and Wanjari, R.H. (2000). Three decades of all India Coordinated Research Project on Long term Fertilizer experiments to study changes in soil quality, crop productivity and sustainability. IISS, Bhopal.

Szabolcs, I. (1969). The role of salt balances in the prediction and prevention of the secondary salinizations and alkalization of irrigated soils.(H) *Nemzetkozi Mezogazd, Tzemle*, **13** : 45-50. (English summary).

Tabatabai, M.A. and Bremner, J.M. (1970). A simple tubidemetric method of determining total sulphur in plant materials, *Agron. J.,* 62 : 805-806.

Takkar, P.N. and Randhawa, N.S. (1978). Micronutrients in agriculture. *Fert. News*, **23**: 3-36.

Talati, N.R., Mathur, G.S. and Attri, S.C. (1975). Characterization of alkali hazard in flood plain soils of Rajasthan Canal area. *J. Indian Soc. Soil. Sci.,* **23:** 410-415.

Tandon H.L.S. (1991). Micronutrients in Indian Agriculture. FDCO. New Delhi. pp. 122.

Thiele G., K. Theisen, M. Bonierbale, T. Walker (2010). Targeting the poor and hungry with potato science. *Potato Journal* 37(3-4): 75-86.

Tisdale S.L., Nelson W.L. and Beaton J.D. (1985). Soil Fertility and Fertilizers, 4th edition, Mc Millan, NY.

Tiwari, K.N.(2002). Nutrient management – Issues and Strategies. *Fertilizer news,* **47(8):** 23-49.

Tosh S., Choudhuri, M.A. and Chatterjee S.K. (1980). Effects of some micronutrients on growth, development and fruit yield of Abelmoschus esculentus (L) Moench. *Sci. Culture.* 46(7): 271-274.

Trehan S.P., Upadhayay N.C., Sud K.C., Kumar Manoj, Jatav M.K. and Lal S.S. (2008). Nutrient Management in Potato, Technical Bulletin No. 90. Central Potato Research Institute, Shimla 64.

Troug, E. (1960). Fifty years of soil testing. Trans. *7th Int. Cong. Soil Sci.* 3: 46-57.

Uehara, G. (1989). Technology transfer in the tropics *Outlook on agriculture* 18: 38-42.

Verma T.S., Thakur P.C. and Singh S. (1997). Effect of bio-fertilizers on vegetable and seed yield of cabbage. *Vegetable Science,* **24**: 1-3.

Walkley, A. and Black, C.A. (1934). An examination of different methods for determining soil organic matter and a proposed modification of the chromic acid titration method. *Soil Sci.,* 37 : 29-38.

Water D. and Coltman R. (1989). Responses of lettuce to pre and post-transplant phosphorus and pre-transplant inoculation with a VA-mycorrhizal fungus. *Plant and Soil* **177**: 151-156.

Yadav, B.K. (2011). Micronutrient Status of Soils under Legume Crops in Arid Region of Western Rajasthan, India. *Academic Journal of Plant Sciences* 4(3): 94-97.

Yadav, R.L. and Meena, M.C. (2009). Available micronutrients status and relationship with soil properties of Degana soil series of Rajasthan. *Journal of the Indian Society of Soil Science.* 57(1): 90-92.

Appendices

Appendix-I

Strength of commonly used acids and alkalis

Reagent/ chemical	Normality (approx.)	Molarity (approx.)	Formula weight	% by weight (approx.)	Specific gravity (approx.)	Millilitres required for 1N/litre solution (approx.)	Millilitres required for 1M/litre solution
Nitric acid	16.0	16.0	63.0	70	1.42	63.7	63.7
Sulphuric acid	35.0	17.5	98.0	98	1.84	28.0	56.0
Hydrochloric acid	11.6	11.6	36.5	37	1.19	82.6	82.6
Phosphoric acid	45.0	15.0	98.0	85	1.71	22.7	68.1
Perchloric acid	10.5	10.5	100.5	65	1.60	108.7	108.7
Ammonium hydroxide	15.0	15.0	35.0	28	0.90	67.6	67.6

Appendix-II

Equivalent and molecular weights of some important compounds needed in chemical analysis

Compound	Formula	Molecular weight (g)	Equivalent weight (g)
Ammonium acetate	CH_3COONH_4	77.08	77.08
Ammonium chloride	NH_4Cl	53.49	53.49
Ammonium fluoride	NH_4F	37.04	37.04
Ammonium nitrate	NH_4NO_3	80.04	80.04
Barium chloride	$BaCl_2.2H_2O$	244.28	122.14
Boric acid	H_3BO_3	61.83	20.61
Calcium acetate	$(CH_3COO)_2Ca$	158.00	79.00
Calcium carbonate	$CaCO_3$	100.09	50.05
Calcium chloride (dihydrate)	$CaCl_2.2H_2O$	147.02	73.51
Calcium hydroxide	$Ca(OH)_2$	74.00	37.00
Calcium nitrate	$Ca(NO_3)_2$	164.00	82.00
Calcium sulphate	$CaSO_4.2H_2O$	172.17	86.08
Ferrous ammonium sulphate	$FeSO_4.(NH_4)_2SO_4.H_2O$	392.13	392.13
Ferrous sulphate	$FeSO_4.7H_2O$	278.01	139.00
Magnesium chloride	$MgCl_2.6H_2O$	203.30	101.65
Magnesium nitrate	$Mg(NO_3)_2.6H_2O$	256.41	128.20
Potassium chloride	KCl	74.55	74.55
Potassium dichromate	$K_2Cr_2O_7$	294.19	49.04
Potassium hydroxide	KOH	56.10	56.10
Potassium permanganate	$KMnO_4$	158.03	31.60
Potassium nitrate	KNO_3	101.10	101.10
Potassium sulphate	K_2SO_4	174.27	87.13
Potassium hydrogen phthalate	$COOH\ C_6H_4\ COOK$	204.22	204.22
Oxalic acid	$C_2H_2O_4.2H_2O$	126.00	63.00
Silver nitrate	$AgNO_3$	169.87	169.87
Sodium acetate (anhydrous)	CH_3COONa	82.04	82.04
Sodium bicarbonate	$NaHCO_3$	84.01	84.01
Sodium carbonate	Na_2CO_3	106.00	53.00
Sodium chloride	$NaCl$	58.45	58.45
Sodium hydroxide	$NaOH$	40.00	40.00
Sodium nitrate	$NaNO_3$	84.99	84.99
Sodium oxalate	$Na_2C_2O_4$	134.00	67.00
Sodium sulphate	Na_2SO_4	142.04	71.02
Sodium thiosulphate	$Na_2S_2O_3.5H_2O$	248.18	248.18

Appendix-IIIA

Units and conversion factors

To convert unit in Column 1 to unit in Column 2 multiply by:	Column 1: SI unit	Column 2: non-SI unit	To convert unit in Column 2 to unit in Column 1 multiply by:
		Area	
2.47	hectare, ha	acre	0.405
247	square kilometre, km^2 $(10^3\ m)^2$	acre	4.05×10^{-3}
0.386	square kilometre, km^2 $(10^3\ m)^2$	square mile, mi^2	2.590
2.47×10^{-4}	square metre, m^2	acre	4.05×10^3
10.76	square metre, m^2	square foot, ft^2	9.29×10^{-2}
1.55×10^{-3}	square millimetre, mm^2 $(10^{-3}\ m)^2$	square inch, in^2	645
		Volume	
9.73×10^{-3}	cubic metre, m^3	acre-inch	102.8
35.3	cubic metre, m^3	cubic foot, ft^3	2.83×10^{-2}
6.10×10^4	cubic metre, m^3	cubic inch, in^3	1.64×10^{-5}
1.057	litre, L $(10^{-3}\ m^3)$	quart (liquid), qt	0.946
3.53×10^{-2}	litre, L $(10^{-3}\ m^3)$	cubic foot, ft^3	28.3
0.265	litre, L $(10^{-3}\ m^3)$	gallon	3.78
33.78	litre, L $(10^{-3}\ m^3)$	pint (fluid), pt	0.473
		Mass	
2.20×10^{-3}	gram, g $(10^{-3}\ kg)$	pound, lb	454
3.52×10^{-2}	gram, g $(10^{-3}\ kg)$	ounce (avdp), oz	28.4
2.205	kilogram, kg	pound, lb	0.454
0.01	kilogram, kg	quintal (metric), q	100
1.10×10^{-3}	kilogram, kg	ton (2 000 lb), ton	907
1.102	megagram, Mg (tonne)	ton (U.S.), ton	0.907
1.102	tonne, t	ton (U.S.), ton	0.907
		Length	
0.621	kilometre, km $(10^3\ m)$	mile	1.609
1.094	metre, m	yard, yd	0.914
3.28	metre, m	foot, ft	0.304
3.94×10^{-2}	millimetre, mm $(10^{-3}\ m)$	inch, in	25.4
10	nanometre, nm $(19^{-9}\ m)$	angstrom, å	0.1
		Yield and rate	
0.893	kilogram per hectare, kg ha^{-1}	pound per acre, lb $acre^{-1}$	1.12
7.77×10^{-2}	kilogram per cubic metre, kg m^{-3}	pound per bushel, lb bu^{-1}	12.87

Appendix-IIIB

Units and conversion factors

To convert unit in Column 1 to unit in Column 2 multiply by:	Column 1: SI unit	Column 2: non-SI unit	To convert unit in Column 2 to unit in Column 1 multiply by:
1.49×10^{-2}	kilogram per hectare, kg ha^{-1}	bushel per acre, 60 lb	67.19
1.59×10^{-2}	kilogram per hectare, kg ha^{-1}	bushel per acre, 56 lb	62.71
1.86×10^{-2}	kilogram per hectare, kg ha^{-1}	bushel per acre, 48 lb	53.75
0.107	litre per hectare, L ha^{-1}	gallon per acre	9.35
893	tonne per hectare, t ha^{-1}	pound per acre, lb $acre^{-1}$	1.12×10^{-3}
893	megagram per hectare, Mg ha^{-1}	pound per acre, lb $acre^{-1}$	1.12×10^{-3}
0.446	megagram per hectare, Mg ha^{-1}	ton (2 000 lb) per acre, ton $acre^{-1}$	2.24
2.24	metre per second, m s^{-1}	mile per hour	0.447
	Pressure		
9.90	megaPascal, MPa (10^6 Pa)	atmosphere	0.101
10	megaPascal, MPa (10^6 Pa)	bar	0.1
1.00	megagram per cubic metre, Mg m^{-3}	gram per cubic centimetre, g cm^{-3}	1.00
2.09×10^{-2}	Pascal, Pa	pound per square foot, lb ft^{-2}	47.9
1.45×10^{-4}	Pascal, Pa	pound per square inch, lb in^{-2}	6.90×10^3
	Temperature		
1.00 (K - 273)	Kelvin, K	Celsius, °C	1.00 (°C +273)
(9/5 °C) + 32	Celsius, °C	Fahrenheit, °F	5/9 (°F - 32)
	Water measurement		
9.73×10^{-3}	cubic metre, m^3	acre-inch, acre-in	102.8
9.81×10^{-3}	cubic metre per hour, m^3 h^{-1}	cubic foot per second, ft^3 s^{-1}	101.9
4.40	cubic metre per hour, m^3 h^{-1}	U.S. gallons per minute, gal min^{-1}	0.227
8.11	hectare-metre, ha-m	acre-foot, acre-ft	0.123
97.28	hectare-metre, ha-m	acre-inch, acre-in	1.03×10^{-2}
8.1×10^{-2}	hectare-centimetre, ha-cm	acre-foot, acre-ft	12.33
	Concentration		
1	centimole per kilogram, cmol kg^{-1} (ion exchange capacity)	milli-equivalents per 100 grams, me 100 g^{-1}	1
0.1	gram per kilogram, g kg^{-1}	percent, %	10
1	milligram per kilogram, mg kg^{-1}	parts per million, ppm	1
	Plant nutrient conversion		
	Element	Oxide	
2.29	P	P_2O_5	0.437
1.20	K	K_2O	0.830
1.39	Ca	CaO	0.715
1.66	Mg	MgO	0.602
1.216	N	NH_3	0.777
4.426	N	NO_3	0.226
6.25	N	Protein	0.160
3.00	S	SO_4	0.330
2.5	S	SO_3	0.440
1.724	Organic C	Organic matter	0.580

Appendix-IV A

Acid forming fertilizers

Name of the fertilizer	Equivalent acidity in Kg for 100 Kg of fertilizer
Ammonium sulphate	110
Ammonium sulphate nitrate	93
Urea	80
Ammonium chloride	128
Ammonium nitrate	60

Appendix-IV B

Equivalent acidity: A term used to express the calcium carbonate equivalent of acidic residue left by a fertilizer material.

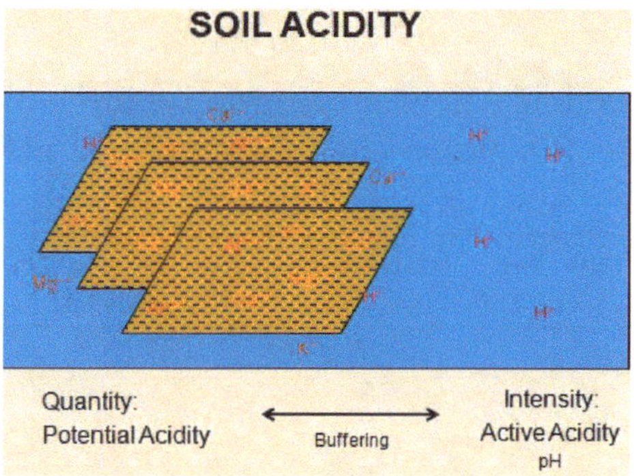

Appendix-IV C

Calcium carbonate equivalent (CCE)

Material	% CCE
$CaCO_3$	100
$MgCO_3$	119
$(Ca,Mg)CO_3$	109
CaO	179
$Ca(OH)_2$	136
$CaSiO_3$	86
Common Limestone	80-109

Appendix-IV D

Alkali forming fertilizers

Name of the fertilizer	Equivalent basicity in Kg for 100 Kg of fertilizer
Sodium nitrate	29
calcium nitrate	21
Calcium cyanamide	63
Dicalcium phosphate	25

Equivalent basicity: A term used to express the calcium carbonate equivalent of basic residue left by a fertilizer material.

Appendix-V

Major properties of selected silicate clay minerals and humus

Properties	Smectite	Vermiculite	Fine mica	Chlorite	Kaolinite	Humas
Size (μm)	0.01-1.0	0.1-5.0	0.2-2.0	0.1-2.0	0.5-5.0	0.1-1.0
Shape	Flakes	Plates; Flakes	Flakes	Variable	Hexagonal crystals	Variable
External surface (m^2/g)	70-120	50-100	70-100	70-100	10-30	500-800
Internal surface (m^2/g)	550-650	500-600	-	70-100	10-30	500-800
Interlayer spacing[b] (nm)	1.0-2.0	1.0-1.5	1.0	1.4	0.7	-
Net negative Charge (cmol/kg)[c]	80-120	100-180	15-40	15-40	2-5	200-750

Appendix-VI A

Charge characteristics of representative colloids showing comparative levels of permanent (constant) and pH dependent negative charges as well as pH dependent positive charges

Colloid type	Total (Cmol/kg)	Constant (%)	Variables (%)	Positive charges (Cmol/kg)
Organic	200	10	90	0
Smectite	100	95	5	0
Vermiculite	150	95	5	0
Fine grained micas	30	80	20	0
Chlorite	30	80	20	0
Kaolinite	8	5	95	2
Gibbsite(AI)	4	0	100	5
Geothite (Fe)	4	0	100	5

Appendix-VI B

Nutrient analysis (per cent on dry weight basis) of different organic preparations

S.No.	Compost	N (%)	P (%)	K (%)	Ca (%)	Zn (ppm)	Cu (ppm)	Fe (ppm)	Mn (ppm)
1	Vermi compost	2.15	1.29	0.53	1.72	168	61	3545	252
2	B D Compost	1.68	0.17	1.23	1.20	96	45	357	3352
3	NADEP Compost	0.98	0.35	1.00	1.25	162	56	430	230
4	MM Compost	1.54	0.51	1.06	1.35	140	45	433	275
5	Farm Yard Manure (FYM)	0.70	0.19	0.37	0.24	75	34	222	235

Appendix-VII

Ionic Radii of elements common in silicate clays and an indication of which are found in the tetrahedral and octahedral sheets

Ion	Radius (nm)[2]	Found in
Si^{4+}	0.042	
Al^{3+}	0.051	Silica tetra hedra
Fe^{3+}	0.064	
Mg^{2+}	0.066	
Zn^{2+}	0.074	Alumina octa hedra
Fe^{2+}	0.070	
Na^{1+}	0.097	Exchange sites
Ca^{2+}	0.099	
K^{+}	1.133	
O^{2-}	0.140	Both sheets
OH^{-}	0.155	

*1 nm $=10^{9}$

Appendix-VIII

All India Demand Forecast of Fertilizer Products ('000 tonnes) 2015-16 to 2019-20

(In '000 t onnes)

Year	Urea	DAP	NP/NPKs	SSP	MOP*
2015-16	32858	12212	11142	5513	4643
2016-17	33677	12413	11420	5948	4793
2017-18	33754	12764	11841	6476	4934
2018-19	34536	12950	12318	6626	5048
2019-20	35307	13014	12799	4096	5086

1. *For direct consumption.
2. Source: Working Group Report on Fertilizer Industry for Twelfth Five-Year Plan

Source: Indian fertilizer scenario 2015

Appendix-IX

Production of Urea, DAP and Complexes

(In Lakh MT)

Year	2010-11	2011-12	2012-13	2013-14	2014-15
Urea	218.81	219.84	225.75	227.15	225.85
DAP	35.37	39.63	36.47	36.11	34.44
Complex Fertilizers	87.27	77.70	61.80	69.13	78.32

Source: Indian fertilizer scenario 2015

Appendix- X

Sector-Wise Production of Complex Fertilizers From 2010-11 to 2014-15

(In '000 MT)

Sector	Annual installed capacity	2010-11	2011-12	2012-13	2013-14	2014-15
Public Sector	2163.5	1247.7	1302.1	1248.4	1220.0	1334.2
Coop. Sector	1635.4	3201.6	2457.6	1433.8	1760.8	1928.0
Private Sector	2042.5	4277.7	4010.5	3498.0	3932.1	4570.2
Total	5841.4	8727.0	7770.2	6180.2	6912.9	7832.4

Source: Indian fertilizer scenario 2015

Appendix-XI

Import of Major Fertilizers

(In LMT)

Year	Urea	DAP	MOP
2010-11	66.10	76.97	63.57
2011-12	78.34	69.05	39.85
2012-13	80.44	57.02	24.96
2013-14	70.88	32.61	31.77
2014-15	87.49	38.17	41.76

Source: Indian fertilizer scenario 2015

Appendix-XII

Production, Import and Consumption of Fertilizers

		2005-06	2006-07	2007-08	2008-09	2009-10	2010-11	2011-12	2012-13	2013-14	2014-15
	1	2	3	4	5	6	7	8	9	10	11
A	**Nitrogenous fertilizers**										
	Production	11354	11578	10899.9	10870	11900	12157	12259	12194	12378	12394
	Imports	1385	2689	3706.5	3751	3447	4493	5240	3505	3920	4766
	Consumption	12723	13773	14419	15090	15580	16558	17300	16820	16760	16946
B	**Phosphatic fertilizers**										
	Production	4221	4517	3806.7	3464	4321	4223	4104	3541	3714	3875
	Imports	1122	1322	1391	3067	2756	3802	4427	2625	1688	1633
	Consumption	5204	5543	5515	6506	7274	8050	7914	6653	6633	6096
C	**Potassic fertilizers**										
	Imports	2747	2069	2653	3403	2945	4069	3335	1178	1926	2537
	Consumption	2413	2335	2636	3313	3632	3514	2676	2062	2099	2532
D	**All Fertilizers**										
	Productoin	15575	16095	14706.5	14334	16221	16380	16363	15735	16092	16269
	Imports	5254	6080	7750.16	10221	9149	12364	13002	7308	7434	9136
	Consumption	20340	21651	22570	24909	26486	28122	27790	25536	24452	25576

Source: Indian fertilizer scenario 2015

Appendix-XIII

Consumption of Fertilizers (State-wise, Nutrient-wise) During 2011-12 to 2014-15

Per Hectare Fertilizer Consumption Kg '(000 Tonnes)

S. No.	State/Ut	2011-12				2012-13				2013-14				2014-15			
		N	P	K	Total	N	P	K	Total	N	P	K	Total	N	P	K	Total
1.	Andhra Pradesh	143.71	75.81	23.41	242.92	130.47	52.3	18.48	201.25	140.36	57.03	22.74	220.13	146.88	60.52	29.83	237.23
2.	Telangana									155.47	45.52	12.00	212.99	162.52	52.71	16.20	231.43
3.	Karnataka	100.83	65.24	27.60	193.68	75.90	33.17	21.27	130.34	88.51	41.41	23.73	153.65	96.31	47.89	91.95	176.15
4.	Kerla	50.92	24.85	37.43	113.20	47.51	24.30	35.13	106.93	26.36	13.64	22.48	62.49	20.23	8.35	11.97	40.55
5.	Tamil Nadu	116.22	54.72	44.81	214.76	111.79	43.39	29.02	184.21	86.77	31.13	25.15	143.05	96.98	36.01	30.68	163.67
6.	Panducherry	541.11	144.44	113.33	798.89	436.15	108.46	62.69	607.31	491.09	84.62	61.94	637.65	196.11	42.57	46.01	284.69
7.	A&N Islands	20.56	20.56	9.44	50.56	17.20	12.80	10.00	40.00	11.77	8.96	7.42	28.15	12.50	9.95	6.63	29.09
8.	Lakshadweep	0.00	0.00	0.00	0.00	0.0	0.0	0.0	0.0	0.0	0.0	0.0	0.0	0.0	0.0	0.0	0.0
	SZ Total	117.04	64.40	29.69	211.13	111.93	41.97	22.43	166.33	102.60	39.87	21.57	164.05	105.63	43.15	25.90	174.68
9.	Gujarat	90.42	31.87	10.14	132.44	79.98	20.40	6.07	106.15	88.60	24.11	6.93	119.63	96.00	27.76	9.03	132.79
10.	Madhya Pradesh	47.16	33.34	3.53	83.03	46.81	30.94	3.07	80.82	51.23	26.85	2.38	80.47	48.68	26.35	3.14	78.17
11.	Chhattisgarh	62.92	31.31	10.92	105.15	65.93	31.63	8.17	105.73	54.41	24.77	7.27	86.45	56.75	26.69	8.63	92.07
12.	Maharashtra	73.19	45.97	18.15	137.30	62.55	32.72	17.76	113.04	67.08	32.69	17.95	117.71	68.76	33.95	22.88	125.60
13.	Rajasthan	37.28	16.98	1.07	55.33	40.35	14.87	0.90	56.12	35.05	10.62	0.19	45.87	39.97	13.91	0.64	54.51
14.	Goa	18.66	16.28	11.52	46.46	17.06	10.00	5.95	33.01	30.66	13.54	9.33	53.53	26.12	11.82	8.91	46.85
15.	Daman & Diu	20.00	3.33	0.00	23.33	46.67	6.67	3.33	56.67	19.05	4.76	2.38	26.19	0.00	0.00	0.00	0.00
16.	D&N Haveli	24.00	16.40	0.00	40.40	24.17	15.83	0.00	40.00	0.0	0.0	0.0	0.0	40.25	27.67	0.00	67.92
	WZ Total	58.31	31.62	7.98	97.84	54.94	25.47	6.92	87.33	56.07	23.18	6.68	85.93	58.76	25.14	8.71	92.61
17.	Haryana	157.33	56.96	5.78	220.07	160.44	48.71	2.61	211.76	168.87	35.26	2.78	206.91	172.18	43.10	6.15	221.44
18.	Punjab	179.20	56.76	6.69	242.64	188.78	58.76	3.06	250.60	174.68	41.65	3.08	219.41	179.03	43.46	4.97	227.46

19.	Uttar Pradesh	118.24	39.49	6.42	164.14	129.81	45.17	5.14	180.12	107.08	27.54	3.77	138.40	111.61	32.24	6.60	150.45
20.	Uttarkhand	109.35	28.54	9.13	147.02	107.91	21.16	6.62	135.69	135.70	21.96	6.47	164.13	128.96	24.11	6.88	159.95
21.	Himachal Pradesh	34.89	10.32	9.50	54.71	35.61	7.20	7.53	50.34	34.08	8.45	8.79	51.32	35.63	8.65	9.67	53.96
22.	J&K	56.96	24.68	4.56	86.19	65.26	20.74	8.39	94.39	0.0	0.0	0.0	0.0	0.0	0.0	0.0	0.0
23.	Delhi	8.97	5.64	0.26	14.87	23.06	1.43	0.0	24.49	32.96	0.00	0.00	32.96	82.25	0.00	0.00	82.25
24.	Chandigarh	0.0	0.0	0.0	0.0	0.0	0.0	0.0	0.0	0.0	0.0	0.0	0.0	0.0	0.0	0.0	0.0
	NZ Total	131.34	43.88	6.45	181.67	140.54	46.00	4.56	191.11	127.68	31.05	3.94	162.67	131.42	35.50	6.66	173.58
25.	Bihar	126.56	38.84	15.09	180.48	142.86	41.86	11.57	196.30	128.67	28.27	12.93	169.88	132.70	31.20	14.75	178.65
26.	Jharkhand	82.53	29.38	7.93	119.84	67.49	45.91	5.86	119.26	30.05	6.31	1.34	37.70	30.75	6.56	1.09	38.39
27.	Odisha	65.15	27.29	11.24	103.68	62.15	24.50	10.06	96.71	36.69	13.80	6.62	57.11	36.14	14.53	6.82	57.49
28.	West Bengal	88.95	50.91	33.04	172.91	85.14	46.88	29.21	161.23	71.65	27.19	28.02	126.86	77.86	40.21	32.84	150.91
	EZ Total	95.80	40.64	21.01	157.45	97.68	40.51	17.92	156.11	72.03	21.43	15.05	108.51	74.75	26.68	17.04	118.47
29.	Assam	36.19	11.76	18.09	66.04	35.99	11.69	17.99	65.68	49.18	13.56	26.04	88.77	68.09	21.91	35.08	125.08
30.	Tripura	28.09	14.80	7.36	50.24	37.17	19.65	12.31	69.13	23.58	17.45	9.62	50.66	18.45	14.25	8.45	41.15
31.	Manipur	18.05	2.66	1.21	21.92	29.39	4.14	1.55	35.08	33.85	5.61	4.75	44.20	45.10	8.80	7.94	61.84
32.	Meghalaya	9.65	3.66	0.74	14.04	9.88	3.38	1.00	14.26	34.41	10.20	2.65	47.25	0.00	0.00	0.00	0.00
33.	Nagaland	1.58	1.03	0.42	3.04	2.25	1.41	0.78	4.44	2.62	1.72	0.83	5.18	3.24	1.93	1.17	6.34
34.	Arunachal Pradesh	1.96	0.36	0.11	2.42	1.58	0.11	0.32	2.00	0.00	0.00	0.00	0.00	0.00	0.00	0.00	0.00
35.	Mizoram	9.48	2.16	0.62	12.27	0.78	0.28	15.17	0.0	0.0	0.0	0.0	0.0	0.0	0.0	0.0	0.0
36.	Sikkim	0.00	0.00	0.00	0.00	0.00	0.00	0.00	0.00	0.0	0.0	0.0	0.0	0.0	0.0	0.0	0.0
	NE Total	27.82	9.23	12.70	49.75	28.87	9.53	13.02	51.42	38.06	11.38	18.52	67.92	45.20	15.27	22.16	82.63
	ALL INDIA	88.43	40.46	13.16	142.05	86.53	34.23	10.61	131.36	81.11	27.28	10.16	118.55	84.86	30.54	12.68	128.08

Source: Indian fertilizer scenario 2015

Appendix-XIV

Consumption Ratio (NPK Consumption Since 1990-91)

S.No.	Year	NPK Ratio
1.	1990-91	6.0 : 2.4 : 1
2.	1991-92	5.9 : 2.4 : 1
3.	1992-93	9.5 : 3.2 : 1
4.	1993-94	9.7 : 2.9 : 1
5.	1994-95	8.4 : 2.6 : 1
6.	1995-96	8.5 : 2.5 : 1
7.	1996-97	10.0 : 2.9 : 1
8.	1997-98	7.9 : 2.8 : 1
9.	1998-99	8.5 : 3.1 : 1
10.	1999-00	6.9 : 2.9 : 1
11.	2000-01	7.0 : 2.7 : 1
12.	2001-02	6.8 : 2.6 : 1
13.	2002-03	6.5 : 2.5 : 1
14.	2003-04	6.9 : 2.6 : 1
15.	2004-05	5.7 : 2.2 : 1
16.	2005-06	5.3 : 2.2 : 1
17.	2006-07	5.9 : 2.4 : 1
18.	2007-08	5.5 : 2.1 : 1
19.	2008-09	4.6 : 2.0 : 1
20.	2009-10	4.3 : 2.0 : 1
21.	2010-11	4.7 : 2.3 : 1
22.	2011-12	6.7 : 3.1 : 1
23.	2012-13	8.2 : 3.2 : 1
24.	2013-14	9.7 : 2.7 : 1
25.	2014-15	6.7 : 2.4 : 1

Source: Indian fertilizer scenario 2015

Appendix-XV

Production of Bio-fertilizer in India

(In Tonnes)

S.No.	Name of the State	Year				
		2010-11	2011-12	2012-13	2013-14	2014-15
South Zone						
1.	Andhra Pradesh	999.00	1125.35	1335.74	2714.22	2688.80
2.	Karnataka	8980.00	5760.32	7683.72	9907.337	16462.62
3.	Kerala	3257.00	904.17	1045.64	3520.66	4916.97
4.	Puducherry	783.00	509.45	621.00	516.98	560.95
5.	Tamil Nadu	8891.00	3373.81	11575.70	14104.83	15373.29
	Total	20550.6	11574.1	22251.80	30764.027	39982.53
West Zone						
6.	Chhattisgarh	0.00	276.34	501.63	712.07	1024.68
7.	Gujarat	6318.00	2037.35	978.48	64114.34	3667.929
8.	Goa	443.40	0.00	370.00	66.29	802.52
9.	Madhya Pradesh	2455.57	2309.06	1408.08	4824.194	2837.99
10.	Maharashtra	2924.00	8743.69	5897.91	6218.607	14847.397
11.	Rajasthan	819.75	199.78	982.00	1315.00	599.898
	Total	12950.72	13595.22	10138.10	19547.565	21580.414
North Zone						
12.	Delhi	1205.00	1617.00	0.00	396.00	104.50
13.	Chandigarh	6.53	914.41	5832.61	1146.483	872.956
14.	Haryana	9.00	1.29	0.00	26.147	0.768
15.	Himachal Pradesh	0.00	0.00	0.00	45.26	0.00
16.	Punjab	2.50	392.22	2311.33	2124.852	6305.453
17.	Uttar Pradesh	1217.45	8695.08	1310.02	2682.221	40.99088
18.	Uttarakhand	45.00	263.01	2758.21	6493.851	2129.952
	Total	2455.48	12183.01	12212.17	11914.814	13512.596
East Zone						
19.	Bihar	136.26	75.00	52.40	52.4	64.90
20.	Jharkhand	0.00	8.36	35.30	74.2	9.08
21.	Odisha	357.65	590.12	407.10	1097.61	1074.48
22.	West Bengal	393.39	603.20	110.00	1682.707	2061.83
	Total	887.31	1276.70	1604.8	2846.917	3210.27
North East Zone						
23.	Arunachal Pradesh	0.00	0.00	0.00	59.0	59.0
24.	Assam	130.00	68.33	89.00	149.0	88.0
25.	Mizoram	2.00	0.00	0.00	4.00	3.60
26.	Nagaland	21.50	13.00	7.45	7.45	7.45
27.	Sikkim	0.00	0.00	9.50	10.1	12.40
28.	Tripura	850.00	1542.85	514.00	225.00	240.0
	Total	1003.5	1624.18	619.95	454.55	410.45
	Grand Total	37997.51	40324.21	46836.82	65527.57	89695.46

Source: Indian fertilizer scenario 2015

Appendix-XVI

Consumption of Fertilizersin Selected Countries in 2011 & 2012

Country	Nitrogen (N)		Phosphate (P_2O_5)		Potash (K_2O)		Total ($N+P_2O_5+K_2O$)	
	2011	2012	2011	2012	2011	2012	2011	2012
Egypt	1207.0	1086.8	150.0	163.0	57.0	60.0	1414.0	1309.8
Canada	2296.7	2479.1	798.0	831.0	297.6	354.6	3393.3	3864.7
USA	12231.4	12050.3	3945.8	4215.1	4185.8	4461.0	20362.8	20726.4
China	38126.3	44976.0	13557.9	16588.7	7790.2	7031.4	59474.4	68596.1
India	17300.3	16820.9	7914.3	6653.4	2575.2	2061.8	27790.0	25536.2
Pakistan	3208.6	2853.0	633.2	746.9	21.0	20.8	3862.8	3620.7
France	2020.0	2139.5	432.1	495.4	481.2	510.1	2933.3	3145.0
Austria	105.0	106.0	40.0	40.0	32.0	36.0	177.0	182.0

Source: Indian fertilizer scenario 2015

Appendix XVII

Methods of soil, plant and water analysis

Chemical parameter	Method used	Reference
Soil Analysis		
Soil Texture	International pipette method	Piper (1950)
Soil reaction (pH)	Soil pH was determined in 1:2 soil water suspension with the help of glass electrode pH meter	Richards (1954)
Electrical conductivity (EC)	Soil EC was determined in 1:2 soil water suspension by using standard conductivity bridge	Richards (1954)
Organic carbon	Organic carbon was determined by Walkley and Black's wet digestion method	Piper (1950)
Available nutrients		
Nitrogen	By alkaline permanganate method	Subbiah and Asija (1956)
Phosphorus	Extract by 0.5 M $NaHCO_3$ at pH 8.5 and development of colour for colorimetric measurement	Olsen *et al.* (1956)
Potassium	Extract with neutral normal ammonium acetate and estimation by flame photometer	Metson (1956)
Fe, Zn, Mn and Cu	Extract with DTPA and determined by AAS	Lindsay and Norvell (1978)
Water soluble cations (1:2 soil water ratio)		
Ca^{2+} and Mg^{2+}	Versinate titration method	Richards (1954)
Na^+ and K^+	By flame photometer	Richards (1954)
Water soluble anions (1:2 soil water ratio)		
CO_3^{--} and HCO_3^-	Titration with standard H_2SO_4 (Method No. 12)	Richards (1954)
Cl^-	Titration with standard $AgNO_3$ (Method No. 13)	Richards (1954)
SO_4^{--}	Sulphate was determined by modified EDTA method	Moghe *et al.* (1964)
Sodium adsorption ratio (SAR)	Calculated with the help of formula $$SAR = \frac{Na^+}{\sqrt{\frac{Ca^{2+} + Mg^{2+}}{2}}}$$	Moghe *et al.* (1964)
All the cations being expressed as me L^{-1}		
Exchangeable sodium percentage	Calculated with the help of formula $$ESP = \frac{100\ (-0.0126 + 0.01475\ SAR)}{1 + (-0.0126 + 0.01475\ SAR)}$$	Moghe *et al.* (1964)
Soluble sodium percentage	Calculated with the help of formula $$SSP = \frac{\text{Soluble } Na^+}{\text{Total soluble cations}} \times 100$$	Richards (1954)

Salt balance study	Salt balance in irrigated soils was calculated by equation $d = b - [a + (cy/mds) \times 10^{-3}]$ Where, d = Salt leached or acuumulated, g/100 g soil b = Amount of salt at end of observation, g/100 g soil a = Amount of salt at r=start of observation, g/100 soil c = Salt content in rain water, g/L v = Amount of monsoon rains, mm m = Thickness of soil layer, m ds = Bulk density of soil, Mg/m^3	Szabolcs (1969)
Chloride accumulation in soil (me L^{-1})	Calculated with the help of formula $Cls = 4.21\left(\frac{Clw.SP}{R}\right)^2$ Where, Cls = Chloride acculmulatin in soil (me L^{-1}) Clw = Chloride content of irrigation water (me L^{-1}) SP = Saturation percentage of soil R = Mean annual rainfall (mm)	Kovda *et al.* (1967)
Permeability index	Calculated with the help of formula $PI = \frac{Na^+ + \sqrt{HCO_3^-}}{Ca^{++} + Mg^{++} + Na^+} \times 100$ All the ions are being expressed as me L^{-1}	Doneen and Headerson (1960)
Dehydrogenase activity	Colorimetric determination of TPF	Casida *et al.* (1964)
Soil microbial biomass C and soil microbial biomass N	By Chloroform fumigation extraction method	Brookes *et al.* (1985); Vance *et al.* (1987).
Plant analysis		
Nitrogen content in seed and straw	Digestion of the plant samples with sulphuric acid and development of colour with Nesseller's reagent for colorimetric estimation	Snell and Snell (1939)
Phosphorus content in seed and straw	Digestion of the plant samples with di-acid and developed colour with vanado molybdate for colorimetric estimation	Jackson (1967)
Potassium content in seed and straw	Digestion of the plant samples with di-acid and determination by flame photometer	Metson (1956)
Fe, Zn, Mn and Cu content in seed and straw	Digestion of the plant samples with di-acid and determination by AAS	Lindsay and Norvell (1978)
N, P, K, Fe, Zn, Mn & Cu uptake	Uptake of N, P, K, Fe, Zn, Mn and Cu were computed from their content in grain and straw and the yield of grain	

and straw using the following relationship.
Uptake of N, P, K, Fe, Zn,
Mn and Cu (Kg ha^{-1})

$$\frac{\begin{array}{l}\text{Content of N, P, K, Fe, Zn, Mn, Cu} \\ \text{in grains}(\%)\times\text{grain yield}\left(\text{kg ha}^{-1}\right) \\ +\text{Content of N, P, K, Fe, Zn, Mn,} \\ \text{Cu in straw}(\%)\times\text{straw yield}\left(\text{kg ha}^{-1}\right)\end{array}}{100}$$

Analysis of irrigation water

pH	Using glass electrode pH meter	Richards (1954)
Electrical conductivity	Using standard conductivity bridge	Richards (1954)
Ca^{++} and Mg^{++}	Versinate titration method	Richards (1954)
Na^{+} & K^{+}	By flame photometer	Richards (1954)
CO_3^{--} and HCO_3^{-}	Titration with standard H_2SO_4 acid (Method No. 12)	Richards (1954)
Cl^{-}	Titration with standard $AgNO_3$ (Method No. 13)	Richards (1954)
SO_4^{--}	Sulphate was determined by modified EDTA method	Moghe *et al.* (1964)
Residual Sodium Carbonate (RSC)	Calculation by formula $RSC = (CO_3^{--} + HCO_3^{-}) - (Ca^{++} + Mg^{++})$ (me L^{-1})	Richards (1954)
Sodium adsorption ratio (SAR)	Calculated with the help of formula $SAR = \frac{Na^{+}}{\sqrt{\frac{Ca^{++}+Mg^{++}}{2}}}$ All the cations are being expressed as me L^{-1}	Moghe *et al.* (1964)
pHc	Calculation by formula $pHc = (pk2\text{-}pkc) + p\,(Ca^{++} + Mg^{++}) + p\,(Alk)$ Where, pHC : A theoritical calculated pH of irrigation water in contact with lime and in equilibrium with soil CO_2; (pk2-pkc) : obtained from using the sum of $Ca^{++} + Mg^{++} + Na^{+}$ in me L^{-1}; p ($Ca^{++} + Mg^{++}$) : obtaned by using the sum of $Ca^{2+} + Mg^{2+}$ in me L^{-1}; p (alk) : obtained by using the sum of CO^{-}_3 and HCO^{-}_3 in me L^{-1}	Ayers and Westcot (1976)

Adjusted SAR	Calculated with the help of formula $$\text{Adj.SAR} = \frac{Na^{+}}{\sqrt{\frac{Ca^{++} + Mg^{++}}{2}}} \times [1 + (8.4 - pHc)]$$ All the cations are being expressed as me L^{-1}	Ayers and Westcot (1976)
Soluble sodium percentage	Calculation by formula $$SSP = \frac{Na^{+}}{Ca^{++} + Mg^{++} + K^{+} + Na^{+}} \times 100$$	Ayers and Westcot (1976)
	Fertility index = $$\frac{\text{\% of samples testing low} \times 1 + \text{\% of samples testing medium} \times 2 + \text{\% of samples testing high} \times 3}{100}$$	
	Salinity index = $$\frac{\text{\% of samples testing normal} \times 1 + \text{\% of samples testing tending to become saline} \times 2 + \text{\% of samples testing saline} \times 3 + \text{\% of samples testing highly saline} \times 4}{100}$$	
	Sodicity index = $$\frac{\text{\% of samples testing normal} \times 1 + \text{\% of samples testing alkaline} \times 2 + \text{\% of samples testing very alkaline} \times 3 + \text{\% of samples testing highly alkaline} \times 4}{100}$$	

FSC
www.fsc.org
MIX
Papier | Fördert
gute Waldnutzung
FSC® C083411

Zeitfracht Medien GmbH
Ferdinand-Jühlke-Straße 7
99095 Erfurt, Deutschland
produktsicherheit@kolibri360.de